Learning to Rank for Information Retrieval

T0135139

Tie-Yan Liu

Learning to Rank for Information Retrieval

 Springer

Tie-Yan Liu
Microsoft Research Asia
Bldg #2, No. 5, Dan Ling Street
Haidian District
Beijing 100080
People's Republic of China
Tie-Yan.Liu@microsoft.com

ISBN 978-3-642-44124-0   ISBN 978-3-642-14267-3 (eBook)
DOI 10.1007/978-3-642-14267-3
Springer Heidelberg Dordrecht London New York

Cover design: KünkelLopka GmbH

Printed on acid-free paper

Springer is part of Springer Science+Business Media (www.springer.com)

Preface

In recent years, with the fast growth of the World Wide Web and the difficulties in finding desired information, efficient and effective information retrieval systems have become more important than ever, and the search engine has become an essential tool for many people. The ranker, a central component in every search engine, is responsible for the matching between processed queries and indexed documents. Because of its central role, great attention has been paid to the research and development of ranking technologies. In addition, ranking is also pivotal for many other information retrieval applications, such as collaborative filtering, question answering, multimedia retrieval, text summarization, and online advertising. Leveraging machine learning technologies in the ranking process has led to innovative and more effective ranking models, and has also led to the emerging of a new research area named learning to rank.

This new book gives a comprehensive review of the major approaches to learning to rank, i.e., the pointwise, pairwise, and listwise approaches. For each approach, the basic framework, example algorithms, and their theoretical properties are discussed. Then some recent advances in learning to rank that are orthogonal to the three major approaches are introduced, including relational ranking, query-dependent ranking, semi-supervised ranking, and transfer ranking. Next, we introduce the benchmark datasets for the research on learning to rank and discuss some practical issues regarding the application of learning to rank, such as click-through log mining and training data selection/preprocessing. After that several examples that apply learning-to-rank technologies to solve real information retrieval problems are presented. The book is completed by theoretical discussions on guarantees for ranking performance, and the outlook of future research on learning to rank.

This book is written for researchers and graduate students in information retrieval and machine learning. Familiarity of machine learning, probability theory, linear algebra, and optimization would be helpful though not essential as the book includes a self-contained brief introduction to the related knowledge in Chaps. 21 and 22. Because learning to rank is still a fast growing research area, it is impossible to provide a complete list of references. Instead, the aim has been to give references that are representative and hopefully provide entry points into the short but rich literature of learning to rank. This book also provides several promising future research directions on learning to rank, hoping that the readers can be inspired to work on these new topics and contribute to this emerging research area in person.

Beijing   Tie-Yan Liu
People's Republic of China
February 14, 2011

I would like to dedicate this book to my wife and my lovely baby son!

Acknowledgements

I would like to take this opportunity to thank my colleagues and interns at Microsoft Research Asia, who have been working together with me on the topic of learning to rank, including Hang Li, Wei-Ying Ma, Tao Qin, Jun Xu, Yanyan Lan, Yuting Liu, Wei Chen, Xiubo Geng, Fen Xia, Yin He, Jiang Bian, Zhe Cao, Mingfeng Tsai, Wenkui Ding, and Di He. I would also like to thank my external collaborators such as Hongyuan Zha, Olivier Chapelle, Yi Chang, Chengxiang Zhai, Thorsten Joachims, Xu-Dong Zhang, and Liwei Wang. Furthermore, without the support of my family, it would be almost impossible for me to finish the book in such a tight schedule. Here I will present my special thanks to my wife, Jia Cui, and all my family members.

Contents

Part I
Overview of Learning to Rank

In this part, we will give an overview of learning to rank for information retrieval. Specifically, we will first introduce the ranking problem in information retrieval by describing typical scenarios, reviewing popular ranking models, and showing widely used evaluation measures. Then we will depict how to use machine learning technologies to solve the ranking problem, through explaining the concept of learning to rank, enumerating the major approaches to learning to rank, and briefly discussing their strengths and weaknesses.

After reading this part, the readers are expected to know the context of learning to rank, and understand its basic concepts and approaches.

Chapter 1
Introduction

Abstract In this chapter, we give a brief introduction to learning to rank for information retrieval. Specifically, we first introduce the ranking problem by taking document retrieval as an example. Second, conventional ranking models proposed in the literature of information retrieval are reviewed, and widely used evaluation measures for ranking are mentioned. Third, the motivation of using machine learning technology to solve the problem of ranking is given, and existing learning-to-rank algorithms are categorized and briefly depicted.

1.1 Overview

With the fast development of the Web, every one of us is experiencing a flood of information. A study[1] conducted in 2005 estimated the World Wide Web to contain 11.5 billion pages by January 2005. In the same year, Yahoo![2] announced that its search engine index contained more than 19.2 billion documents. It was estimated by http://www.worldwidewebsize.com/ that there were about 25 billion pages indexed by major search engines as of October 2008. Recently, the Google blog[3] reported that about one trillion web pages have been seen during their crawling and indexing. According to the above information, we can see that the number of webpages is growing very fast. Actually, the same story also happens to the number of websites. According to a report,[4] the evolution of websites from 2000 to 2007 is shown in Fig. 1.1.

The extremely large size of the Web makes it generally impossible for common users to locate their desired information by browsing the Web. As a consequence, efficient and effective information retrieval has become more important than ever, and the search engine (or information retrieval system) has become an essential tool for many people.

A typical search engine architecture is shown in Fig. 1.2. As can be seen from the figure, there are in general six major components in a search engine: crawler, parser,

---

[1]http://www.cs.uiowa.edu/~asignori/web-size/.

[2]http://www.iht.com/articles/2005/08/15/business/web.php.

[3]http://googleblog.blogspot.com/2008/07/we-knew-web-was-big.html.

[4]http://therawfeed.com/.

Fig. 1.1 Number of websites

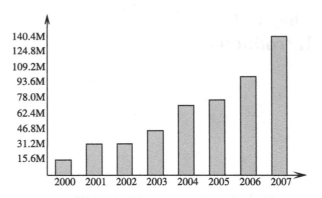

Fig. 1.2 Search engine overview

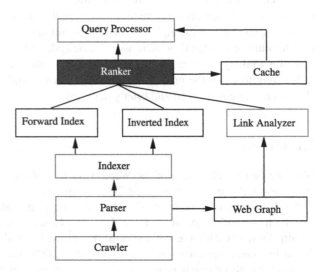

indexer, link analyzer, query processor, and ranker. The crawler collects webpages and other documents from the Web, according to some prioritization strategies. The parser analyzes these documents and generates index terms and a hyperlink graph for them. The indexer takes the output of the parser and creates the indexes or data structures that enable fast search of the documents. The link analyzer takes the Web graph as input, and determines the importance of each page. This importance can be used to prioritize the recrawling of a page, to determine the tiering, and to serve as a feature for ranking. The query processor provides the interface between users and search engines. The input queries are processed (e.g., removing stop words, stemming, etc.) and transformed to index terms that are understandable by search engines. The ranker, which is a central component, is responsible for the matching between processed queries and indexed documents. The ranker can directly take the queries and documents as inputs and compute a matching score using some heuristic formulas, and can also extract some features for each query-document pair and combine these features to produce the matching score.

Partly because of the central role of ranker in search engines, great attention has been paid to the research and development of ranking technologies. Note that ranking is also the central problem in many other information retrieval applications, such as collaborative filtering [30], question answering [3, 71, 79, 83], multimedia retrieval [86, 87], text summarization [53], and online advertising [16, 49]. In this book, we will mainly take document retrieval in search as an example. Note that even document retrieval is not a narrow task. There are many different ranking scenarios of interest for document retrieval. For example, sometimes we need to rank documents purely according to their relevance with regards to the query. In some other cases, we need to consider the relationships of similarity [73], website structure [22], and diversity [91] between documents in the ranking process. This is also referred to as relational ranking [61].

To tackle the problem of document retrieval, many heuristic ranking models have been proposed and used in the literature of information retrieval. Recently, given the amount of potential training data available, it has become possible to leverage machine learning technologies to build effective ranking models. Specifically, we call those methods that learn how to combine predefined features for ranking by means of discriminative learning "learning-to-rank" methods.

In recent years, learning to rank has become a very hot research direction in information retrieval. First, a large number of learning-to-rank papers have been published at top conferences on machine learning and information retrieval. Examples include [4, 7–10, 15, 18, 20, 21, 25, 26, 31, 37, 44, 47, 55, 58, 62, 68, 72, 76, 82, 88, 89]. There have been multiple sessions in recent SIGIR conferences dedicated for the topic on learning to rank. Second, benchmark datasets like LETOR [48] have been released to facilitate the research on learning to rank.[5] Many research papers on learning to rank have used these datasets for their experiments, which make their results easy to compare. Third, several activities regarding learning to rank have been organized. For example, the workshop series on Learning to Rank for Information Retrieval (2007–2009), the workshop on Beyond Binary Relevance: Preferences, Diversity, and Set-Level Judgments (2008), and the workshop on Redundancy, Diversity, and Interdependent Document Relevance (2009) have been organized at SIGIR. A special issue on learning to rank has been organized at Information Retrieval Journal (2009). In Table 1.1, we have listed the major activities regarding learning to rank in recent years. Active participation of researchers in these activities has demonstrated the continued interest from the research community on the topic of learning to rank. Fourth, learning to rank has also become a key technology in the industry. Several major search engine companies are using learning-to-rank technologies to train their ranking models.[6]

When a research area comes to this stage, several questions as follows naturally arise.

---

[5]http://research.microsoft.com/~LETOR/.

[6]http://glinden.blogspot.com/2005/06/msn-search-and-learning-to-rank.html,
http://www.ysearchblog.com/2008/07/09/boss-the-next-step-in-our-open-search-ecosystem/.

Table 1.1 Recent events on learning to rank

Year	Event
2010	SIGIR Tutorial on Learning to Rank for Information Retrieval *Given by Tie-Yan Liu* Learning to Rank Challenge *Organized by Yahoo! Labs* Microsoft Learning to Rank Dataset *Released by Microsoft Research Asia* ICML Workshop on Learning to Rank *Organized by Olivier Chapelle, Yi Chang, and Tie-Yan Liu*
2009	Special Issue on Learning to Rank at Information Retrieval Journal *Edited by Tie-Yan Liu, Thorsten Joachims, Hang Li, and Chengxiang Zhai* NIPS Workshop on Learning with Orderings *Organized by Tiberio Caetano, Carlos Guestrin, Jonathan Huang, Guy Lebanon,* *Risi Kondor, and Marina Meila* NIPS Workshop on Advances in Ranking *Organized by Shivani Agarwal, Chris J.C. Burges, and Koby Crammer* SIGIR Workshop on Redundancy, Diversity, and Interdependent Document Relevance *Organized by Paul N. Bennett, Ben Carterette, Thorsten Joachims, and Filip Radlinski* SIGIR Workshop on Learning to Rank for Information Retrieval *Organized by Hang Li, Tie-Yan Liu, and ChengXiang Zhai* LETOR Dataset 4.0 *Released by Microsoft Research Asia* ACL-IJNLP Tutorial on Learning to Rank *Given by Hang Li* WWW Tutorial on Learning to Rank for Information Retrieval *Given by Tie-Yan Liu* ICML Tutorial on Machine Learning in Information Retrieval: Recent Successes and New Opportunities *Given by Paul Bennett, Misha Bilenko, and Kevyn Collins-Thompson*
2008	SIGIR Workshop on Learning to Rank for Information Retrieval *Organized by Hang Li, Tie-Yan Liu, and ChengXiang Zhai* LETOR Dataset 3.0 *Released by Microsoft Research Asia* SIGIR Tutorial on Learning to Rank for Information Retrieval *Given by Tie-Yan Liu* WWW Tutorial on Learning to Rank for Information Retrieval *Given by Tie-Yan Liu*
2007	SIGIR Workshop on Learning to Rank for Information Retrieval *Organized by Thorsten Joachims, Hang Li, Tie-Yan Liu, and ChengXiang Zhai* WWW Tutorial on Learning to Rank in Vector Spaces and Social Networks *Given by Soumen Chakrabarti* LETOR Dataset 1.0 and 2.0 *Released by Microsoft Research Asia*
2006	ICML Workshop on Learning in Structured Output Space *Organized by Ulf Brefeld, Thorsten Joachims, Ben Taskar, and Eric P. Xing*
2005	NIPS Workshop on Learning to Rank *Organized by Shivani Agarwal, Corinna Cortes, and Ralf Herbrich*

- In what respect are these learning-to-rank algorithms similar and in which aspects do they differ? What are the strengths and weaknesses of each algorithm?
- Empirically speaking, which of those many learning-to-rank algorithms performs the best?
- Theoretically speaking, is ranking a new machine learning problem, or can it be simply reduced to existing machine learning problems? What are the unique theoretical issues for ranking that should be investigated?
- Are there many remaining issues regarding learning to rank to study in the future?

The above questions have been brought to the attention of the information retrieval and machine learning communities in a variety of contexts, especially during recent years. The aim of this book is to answer these questions. Needless to say, the comprehensive understanding of the task of ranking in information retrieval is the first step to find the right answers. Therefore, we will first give a brief review of ranking in information retrieval, and then formalize the problem of learning to rank so as to set the stage for the upcoming detailed discussions.

1.2 Ranking in Information Retrieval

In this section, we will briefly review representative ranking models in the literature of information retrieval, and introduce how these models are evaluated.

1.2.1 Conventional Ranking Models

In the literature of information retrieval, many ranking models have been proposed [2]. They can be roughly categorized as relevance ranking models and importance ranking models.

1.2.1.1 Relevance Ranking Models

The goal of a relevance ranking model is to produce a ranked list of documents according to the relevance between these documents and the query. Although not necessary, for ease of implementation, the relevance ranking model usually takes each individual document as an input, and computes a score measuring the matching between the document and the query. Then all the documents are sorted in descending order of their scores.

The early relevance ranking models retrieve documents based on the occurrences of the query terms in the documents. Examples include the *Boolean model* [2]. Basically these models can predict whether a document is relevant to the query or not, but cannot predict the degree of relevance.

To further model the relevance degree, the *Vector Space model* (VSM) was proposed [2]. Both documents and queries are represented as vectors in a Euclidean space, in which the inner product of two vectors can be used to measure their similarities. To get an effective vector representation of the query and the documents, TF-IDF weighting has been widely used.[7] The TF of a term t in a vector is defined as the normalized number of its occurrences in the document, and the IDF of it is defined as follows:

$$IDF(t) = \log \frac{N}{n(t)} \qquad (1.1)$$

where N is the total number of documents in the collection, and $n(t)$ is the number of documents containing term t.

While VSM implies the assumption on the independence between terms, *Latent Semantic Indexing* (LSI) [23] tries to avoid this assumption. In particular, Singular Value Decomposition (SVD) is used to linearly transform the original feature space to a "latent semantic space". Similarity in this new space is then used to define the relevance between the query and the documents.

As compared with the above, models based on the probabilistic ranking principle [50] have garnered more attention and achieved more success in past decades. The famous ranking models like the *BM25 model*[8] [65] and the *language model for information retrieval (LMIR)*, can both be categorized as probabilistic ranking models.

The basic idea of BM25 is to rank documents by the log-odds of their relevance. Actually BM25 is not a single model, but defines a whole family of ranking models, with slightly different components and parameters. One of the popular instantiations of the model is as follows.

Given a query q, containing terms $t_1, \ldots, t_M$, the BM25 score of a document d is computed as

$$BM25(d, q) = \sum_{i=1}^{M} \frac{IDF(t_i) \cdot TF(t_i, d) \cdot (k_1 + 1)}{TF(t_i, d) + k_1 \cdot (1 - b + b \cdot \frac{LEN(d)}{avdl})}, \qquad (1.2)$$

where $TF(t, d)$ is the term frequency of t in document d, $LEN(d)$ is the length (number of words) of document d, and $avdl$ is the average document length in the text collection from which documents are drawn. k_1 and b are free parameters, $IDF(t)$ is the IDF weight of the term t, computed by using (1.1), for example.

LMIR [57] is an application of the statistical language model on information retrieval. A statistical language model assigns a probability to a sequence of terms. When used in information retrieval, a language model is associated with a document.

---

[7]Note that there are many different definitions of TF and IDF in the literature. Some are purely based on the frequency and the others include smoothing or normalization [70]. Here we just give some simple examples to illustrate the main idea.

[8]The name of the actual model is BM25. In the right context, however, it is usually referred to as "OKapi BM25", since the OKapi system was the first system to implement this model.

With query q as input, documents are ranked based on the query likelihood, or the probability that the document's language model will generate the terms in the query (i.e., $P(q|d)$). By further assuming the independence between terms, one has $P(q|d) = \prod_{i=1}^{M} P(t_i|d)$, if query q contains terms $t_1, \ldots, t_M$.

To learn the document's language model, a maximum likelihood method is used. As in many maximum likelihood methods, the issue of smoothing the estimate is critical. Usually a background language model estimated using the entire collection is used for this purpose. Then, the document's language model can be constructed as follows:

$$p(t_i|d) = (1 - \lambda)\frac{TF(t_i, d)}{LEN(d)} + \lambda p(t_i|C), \qquad (1.3)$$

where $p(t_i|C)$ is the background language model for term t_i, and $\lambda \in [0, 1]$ is a smoothing factor.

There are many variants of LMIR, some of them even go beyond the query likelihood retrieval model (e.g., the models based on K-L divergence [92]). We would not like to introduce more about it, and readers are encouraged to read the tutorial authored by Zhai [90].

In addition to the above examples, many other models have also been proposed to compute the relevance between a query and a document. Some of them [74] take the proximity of the query terms into consideration, and some others consider the relationship between documents in terms of content similarity [73], hyperlink structure [67], website structure [60], and topic diversity [91].

1.2.1.2 Importance Ranking Models

In the literature of information retrieval, there are also many models that rank documents based on their own importance. We will take PageRank as an example for illustration. This model is particularly applicable to Web search because it makes use of the hyperlink structure of the Web for ranking.

PageRank uses the probability that a surfer randomly clicking on links will arrive at a particular webpage to rank the webpages. In the general case, the PageRank value for any page d_u can be expressed as

$$PR(d_u) = \sum_{d_v \in B_u} \frac{PR(d_v)}{U(d_v)}. \qquad (1.4)$$

That is, the PageRank value for a page d_u is dependent on the PageRank values for each page d_v out of the set B_u (containing all pages linking to page d_u), divided by $U(d_v)$, the number of outlinks from page d_v.

To get a meaningful solution to (1.4), a smoothing term is introduced. When a random surfer walks on the link graph, she/he does not necessarily always follow the existing hyperlinks. There is a small probability that she/he will jump to any other

Fig. 1.3 Illustration of
PageRank

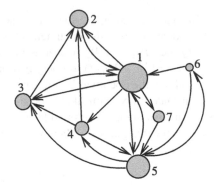

page uniformly. This small probability can be represented by $(1 - \alpha)$, where α is
called the damping factor. Accordingly, the PageRank model is refined as follows[9]:

$$PR(d_u) = \alpha \sum_{d_v \in B_u} \frac{PR(d_v)}{U(d_v)} + \frac{(1 - \alpha)}{N}, \qquad (1.5)$$

where N is the total number of pages on the Web.

The above process of computing PageRank can be vividly illustrated by Fig. 1.3.
By using the above formula, it is not difficult to discover that the PageRank values
of the seven nodes in the graph are 0.304, 0.166, 0.141, 0.105, 0.179, 0.045, and
0.060, respectively. The sizes of the circles in the figure are used to represent the
PageRank values of the nodes.

Many algorithms have been developed in order to further improve the accuracy
and efficiency of PageRank. Some work focuses on the speed-up of the computation
[1, 32, 51], while others focus on the refinement and enrichment of the model. For
example, topic-sensitive PageRank [35] and query-dependent PageRank [64] introduce topics and assume that the endorsement from a page belonging to the same
topic is larger than that from a page belonging to a different topic. Other variations
of PageRank include those modifying the 'personalized vector' [34], changing the
'damping factor' [6], and introducing inter-domain and intra-domain link weights
[46]. Besides, there is also some work on the theoretic issues of PageRank [5, 33].
Langville et al. [46] provide a good survey on PageRank and its related work.

Algorithms that can generate robust importance ranking against link spam have
also been proposed. For example, TrustRank [29] is an importance ranking algorithm that takes into consideration the reliability of web pages when calculating the
importance of pages. In TrustRank, a set of reliable pages are first identified as seed
pages. Then the *trust* of a seed page is propagated to other pages on the web link
graph. Since the propagation in TrustRank starts from reliable pages, TrustRank can
be more spam-resistant than PageRank.

---

[9]If there are web pages without any inlinks (which is usually referred to as dangling nodes in the
graph), some additional heuristics is needed to avoid rank leak.

1.2.2 Query-Level Position-Based Evaluations

Given the large number of ranking models as introduced in the previous subsection, a standard evaluation mechanism is needed to select the most effective model.

Actually, evaluation has played a very important role in the history of information retrieval. Information retrieval is an empirical science; and it has been a leader in computer science in understanding the importance of evaluation and benchmarking. Information retrieval has been well served by the Cranfield experimental methodology [81], which is based on sharable document collections, information needs (queries), and relevance assessments. By applying the Cranfield paradigm to document retrieval, the corresponding evaluation process can be described as follows.

- Collect a large number of (randomly sampled) queries to form a test set.
- For each query q,
    - Collect documents $\{d_j\}_{j=1}^{m}$ associated with the query.
    - Get the relevance judgment for each document by human assessment.
    - Use a given ranking model to rank the documents.
    - Measure the difference between the ranking results and the relevance judgment using an evaluation measure.
- Use the average measure on all the queries in the test set to evaluate the performance of the ranking model.

As for collecting the documents associated with a query, a number of strategies can be used. For example, one can simply collect all the documents containing the query word. One can also choose to use some predefined rankers to get documents that are more likely to be relevant. A popular strategy is the pooling method used in TREC.[10] In this method a pool of possibly relevant documents is created by taking a sample of documents selected by the various participating systems. In particular, the top 100 documents retrieved in each submitted run for a given query are selected and merged into the pool for human assessment.

As for the relevance judgment, three strategies have been used in the literature.

1. *Relevance degree*: Human annotators specify whether a document is relevant or not to the query (i.e., binary judgment), or further specify the degree of relevance (i.e., multiple ordered categories, e.g., Perfect, Excellent, Good, Fair, or Bad). Suppose for document d_j associated with query q, we get its relevance judgment as l_j. Then for two documents d_u and d_v, if $l_u \succ l_v$, we say that document d_u is more relevant than document d_v with regards to query q, according to the relevance judgment.
2. *Pairwise preference*: Human annotators specify whether a document is more relevant than the other with regards to a query. For example, if document d_u is judged to be more relevant than document d_v, we give the judgment $l_{u,v} = 1$;

---

[10]http://trec.nist.gov/.

otherwise, $l_{u,v} = -1$. That is, this kind of judgment captures the relative preference between documents.[11]

3. *Total order*: Human annotators specify the total order of the documents with respect to a query. For the set of documents $\{d_j\}_{j=1}^m$ associated with query q, this kind of judgment is usually represented as a certain permutation of these documents, denoted as π_l.

Among the aforementioned three kinds of judgments, the first kind is the most popularly used judgment. This is partially because this kind of judgment is easy to obtain. Human assessors only need to look at each individual document to produce the judgment. Comparatively, obtaining the third kind of judgment is the most costly. Therefore, in this book, we will mostly use the first kind of judgment as an example to perform the discussions.

Given the vital role that relevance judgments play in a test collection, it is important to assess the quality of the judgments. In previous practices like TREC, both the completeness and the consistency of the relevance judgments are of interest. Completeness measures the degree to which all the relevant documents for a topic have been found; consistency measures the degree to which the assessor has marked all the "truly" relevant documents as relevant and the "truly" irrelevant documents as irrelevant.

Since manual judgment is always time consuming, it is almost impossible to judge all the documents with regards to a query. Consequently, there are always unjudged documents returned by the ranking model. As a common practice, one regards the unjudged documents as irrelevant in the evaluation process.

With the relevance judgment, several evaluation measures have been proposed and used in the literature of information retrieval. It is clear that understanding these measures will be very important for learning to rank, since to some extent they define the "true" objective function of ranking. Below we list some popularly used measures. In order to better understand these measures, we use the example shown in Fig. 1.4 to perform some quantitative calculation with respect to each measure. In the example, there are three documents retrieved for the query "learning to rank", and binary judgment on the relevance of each document is provided.

Most of the evaluation measures are defined first for each query, as a function of the ranked list π given by the ranking model and the relevance judgment. Then the measures are averaged over all the queries in the test set.

As will be seen below, the maximum values for some evaluation measures, such as MRR, MAP, and NDCG are one. Therefore, we can consider one minus these measures (e.g., $(1 - \text{MRR})$, $(1 - \text{NDCG})$, and $(1 - \text{MAP})$) as ranking errors. For ease of reference, we call them *measure-based ranking errors*.

Mean Reciprocal Rank (MRR)   For query q, the rank position of its first relevant document is denoted as r_1. Then $\frac{1}{r_1}$ is defined as MRR for query q. It is clear that documents ranked below r_1 are not considered in MRR.

---

[11]This kind of judgment can also be mined from click-through logs of search engines [41, 42, 63].

Fig. 1.4 Retrieval result for query "learning to rank"

Query = learning to rank

1. http://research.microsoft.com/~letor/	Relevant
2. http://www.learn-in-china.com/rank.htm	Irrelevant
3. http://web.mit.edu/shivani/www/Ranking-NIPS-05/	Relevant
... ...	

Consider the example as shown in Fig. 1.4. Since the first document in the retrieval result is relevant, $r_1 = 1$. Therefore, MRR for this query equals 1.

Mean Average Precision (MAP) To define MAP [2], one needs to define Precision at position k ($P@k$) first. Suppose we have binary judgment for the documents, i.e., the label is one for relevant documents and zero for irrelevant documents. Then $P@k$ is defined as

$$P@k(\pi, l) = \frac{\sum_{t \leq k} I_{\{l_{\pi^{-1}(t)}=1\}}}{k}, \tag{1.6}$$

where $I_{\{\cdot\}}$ is the indicator function, and $\pi^{-1}(j)$ denotes the document ranked at position j of the list π.

Then the Average Precision (AP) is defined by

$$AP(\pi, l) = \frac{\sum_{k=1}^{m} P@k \cdot I_{\{l_{\pi^{-1}(k)}=1\}}}{m_1}, \tag{1.7}$$

where m is the total number of documents associated with query q, and m_1 is the number of documents with label one.

The mean value of AP over all the test queries is called mean average precision (MAP).

Consider the example as shown in Fig. 1.4. Since the first document in the retrieval result is relevant, it is clear $P@1 = 1$. Because the second document is irrelevant, we have $P@2 = \frac{1}{2}$. Then for $P@3$, since the third document is relevant, we obtain $P@3 = \frac{2}{3}$. Then $AP = \frac{1}{2}(1 + \frac{2}{3}) = \frac{5}{6}$.

Discounted Cumulative Gain (DCG) DCG [39, 40] is an evaluation measure that can leverage the relevance judgment in terms of multiple ordered categories, and has an explicit position discount factor in its definition. More formally, suppose the ranked list for query q is π, then DCG at position k is defined as follows:

$$DCG@k(\pi, l) = \sum_{j=1}^{k} G(l_{\pi^{-1}(j)})\eta(j), \tag{1.8}$$

where $G(\cdot)$ is the rating of a document (one usually sets $G(z) = (2^z - 1)$), and $\eta(j)$ is a position discount factor (one usually sets $\eta(j) = 1/\log(j+1)$).

By normalizing DCG@k with its maximum possible value (denoted as Z_k), we will get another measure named Normalized DCG (NDCG). That is,

$$\text{NDCG}@k(\pi, l) = \frac{1}{Z_k} \sum_{j=1}^{k} G(l_{\pi^{-1}(j)})\eta(j). \tag{1.9}$$

It is clear that NDCG takes values from 0 to 1.

If we want to consider all the labeled documents (whose number is m) to compute NDCG, we will get NDCG@m (and sometimes NDCG for short).

Consider the example as shown in Fig. 1.4. It is easy to see that DCG@3 $= 1.5$, and $Z_3 = 1.63$. Correspondingly, NDCG = NDCG@3 $= \frac{1.5}{1.63} = 0.92$.

Rank Correlation (RC) The correlation between the ranked list given by the model (denoted as π) and the relevance judgment (denoted as π_l) can be used as a measure. For example, when the weighted Kendall's τ [43] is used, the rank correlation measures the weighted pairwise inconsistency between two lists. Its definition is given by

$$\tau_K(\pi, \pi_l) = \frac{\sum_{u<v} w_{u,v}(1 + \text{sgn}((\pi(u) - \pi(v))(\pi_l(u) - \pi_l(v))))}{2\sum_{u<v} w_{u,v}}, \tag{1.10}$$

where $w_{u,v}$ is the weight, and $\pi(u)$ means the rank position of document d_u in a permutation π.

Note that the above rank correlation is defined with the assumption that the judgment is given as total order π_l. When the judgments are in other forms, we need to introduce the concept of an equivalent permutation set to generalize the above definition. That is, there will be multiple permutations that are consistent with the judgment in terms of relevance degrees or pairwise preferences. The set of such permutations is called the equivalent permutation set, denoted as Ω_l.

Specifically, for the judgment in terms of relevance degree, the equivalent permutation set is defined as follows.

$$\Omega_l = \{\pi_l | u < v, \text{ if } l_{\pi_l^{-1}(u)} \succ l_{\pi_l^{-1}(v)}\}.$$

Similarly, for the judgment in terms of pairwise preferences, Ω_l is defined as below.

$$\Omega_l = \{\pi_l | u < v, \text{ if } l_{\pi_l^{-1}(u),\pi_l^{-1}(v)} = 1\}.$$

Then, we can refine the definition of weighted Kendall's τ as follows.

$$\tau_K(\pi, \Omega_l) = \max_{\pi_l \in \Omega_l} \tau_K(\pi, \pi_l). \tag{1.11}$$

Consider the example as shown in Fig. 1.4. It is clear that there are multiple permutations consistent with the labels: (1, 3, 2) and (3, 1, 2). Since the ranking result is (1, 2, 3), it is not difficult to compute that the τ_K between (1, 2, 3) and

$(1, 3, 2)$ is $\frac{2}{3}$ and the τ_K between $(1, 2, 3)$ and $(3, 1, 2)$ is $\frac{1}{3}$. Therefore, we can obtain that $\tau_K(\pi, \Omega_l) = \frac{2}{3}$ in this case.

To summarize, there are some common properties in these evaluation measures.[12]

1. All these evaluation measures are calculated at the *query level*. That is, first the measure is computed for each query, and then averaged over all queries in the test set. No matter how poorly the documents associated with a particular query are ranked, it will not dominate the evaluation process since each query contributes similarly to the average measure.

2. All these measures are *position based*. That is, rank position is explicitly used. Considering that with small changes in the scores given by a ranking model, the rank positions will not change until one document's score passes another, the position-based measures are usually discontinuous and non-differentiable with regards to the scores. This makes the optimization of these measures quite difficult. We will conduct more discussions on this in Sect. 4.2.

Note that although when designing ranking models, many researchers have taken the assumption that the ranking models can assign a score to each query-document pair independently of other documents; when performing evaluation, all the documents associated with a query are considered together. Otherwise, one cannot determine the rank position of a document and the aforementioned measures cannot be defined.

1.3 Learning to Rank

Many ranking models have been introduced in the previous section, most of which contain parameters. For example, there are parameters k_1 and b in BM25 (see (1.2)), parameter λ in LMIR (see (1.3)), and parameter α in PageRank (see (1.5)). In order to get a reasonably good ranking performance (in terms of evaluation measures), one needs to tune these parameters using a validation set. Nevertheless, parameter tuning is far from trivial, especially considering that evaluation measures are discontinuous and non-differentiable with respect to the parameters. In addition, a model perfectly tuned on the validation set sometimes performs poorly on unseen test queries. This is usually called over-fitting. Another issue is regarding the combination of these ranking models. Given that many models have been proposed in the literature, it is natural to investigate how to combine these models and create an even more effective new model. This is, however, not straightforward either.

---

[12]Note that this is not a complete introduction of evaluation measures for information retrieval. There are several other measures proposed in the literature, some of which even consider the novelty and diversity in the search results in addition to the relevance. One may want to refer to [2, 17, 56, 91] for more information.

While information retrieval researchers were suffering from these problems, machine learning has been demonstrating its effectiveness in automatically tuning parameters, in combining multiple pieces of evidence, and in avoiding over-fitting. Therefore, it seems quite promising to adopt machine learning technologies to solve the aforementioned problems in ranking.

1.3.1 Machine Learning Framework

In many machine learning researches (especially discriminative learning), attention has been paid to the following key components.[13]

1. The *input space*, which contains the objects under investigation. Usually objects are represented by feature vectors, extracted according to different applications.
2. The *output space*, which contains the learning target with respect to the input objects. There are two related but different definitions of the output space in machine learning.[14] The first is the output space of the task, which is highly dependent on the application. For example, in regression, the output space is the space of real numbers $\mathcal{R}$; in classification it is the set of discrete categories $\{1, 2, \ldots, K\}$. The second is the output space to facilitate the learning process. This may differ from the output space of the task. For example, when one uses regression technologies to solve the problem of classification, the output space that facilitates learning is the space of real numbers but not discrete categories.
3. The *hypothesis* space, which defines the class of functions mapping the input space to the output space. That is, the functions operate on the feature vectors of the input objects, and make predictions according to the format of the output space.
4. In order to learn the optimal hypothesis, a training set is usually used, which contains a number of objects and their ground truth labels, sampled from the product of the input and output spaces. The *loss function* measures to what degree the prediction generated by the hypothesis is in accordance with the ground truth label. For example, widely used loss functions for classification include the exponential loss, the hinge loss, and the logistic loss. It is clear that the loss function plays a central role in machine learning, since it encodes the understanding of the target application (i.e., what prediction is correct and what is not). With the loss function, an empirical risk can be defined on the training set, and the optimal hypothesis is usually (but not always) learned by means of empirical risk minimization.

We plot the relationship between these four components in Fig. 1.5 for ease of understanding.

---

[13]For a more comprehensive introduction to the machine learning literature, please refer to [54].

[14]In this book, when we mention the output space, we mainly refer to the second type.

Fig. 1.5 Machine learning framework

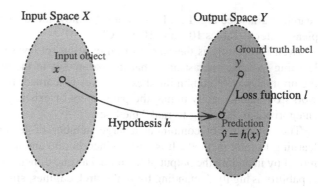

1.3.2 Definition of Learning to Rank

In recent years, more and more machine learning technologies have been used to train the ranking model, and a new research area named "learning to rank" has gradually emerged. Especially in the past several years, learning to rank has become one of the most active research areas in information retrieval.

In general, we call all those methods that use machine learning technologies to solve the problem of ranking "learning-to-rank" methods.[15] Examples include the work on relevance feedback[16] [24, 66] and automatically tuning the parameters of existing information retrieval models [36, 75]. However, most of the state-of-the-art learning-to-rank algorithms learn the optimal way of combining features exacted from query-document pairs through discriminative training. Therefore, in this book we define learning to rank in a more narrow and specific way to better summarize these algorithms. That is, we call those ranking methods that have the following two properties learning-to-rank methods.

Feature Based *"Feature based"* means that all the documents under investigation are represented by feature vectors,[17] reflecting the relevance of the documents to the query. That is, for a given query q, its associated document d can be represented by a vector $x = \Phi(d, q)$, where Φ is a feature extractor. Typical features used in learning to rank include the frequencies of the query terms in the document, the outputs of the BM25 model and the PageRank model, and even the relationship between this document and other documents. These features can be extracted from the index of a

---

[15]In the literature of machine learning, there is a topic named label ranking. It predicts the ranking of multiple class labels for an individual document, but not the ranking of documents. In this regard, it is largely different from the task of ranking for information retrieval.

[16]We will make further discussions on the relationship between relevance feedback and learning to rank in Chap. 2.

[17]Note that, in this book, when we refer to a document, we will not use d any longer. Instead, we will directly use its feature representation x. Furthermore, since our discussions will focus more on the learning process, we will always assume the features are pre-specified, and will not purposely discuss how to extract them.

search engine (see Fig. 1.2). If one wants to know more about widely used features, please refer to Tables 10.2 and 10.3 in Chap. 10.

Even if a feature is the output of an existing retrieval model, in the context of learning to rank, one assumes that the parameter in the model is fixed, and only the optimal way of combining these features is learned. In this sense, the previous works on automatically tuning the parameters of existing models [36, 75] are not categorized as "learning-to-rank" methods.

The capability of combining a large number of features is an advantage of learning-to-rank methods. It is easy to incorporate any new progress on a retrieval model by including the output of the model as one dimension of the features. Such a capability is highly demanding for real search engines, since it is almost impossible to use only a few factors to satisfy the complex information needs of Web users.

Discriminative Training *"Discriminative training"* means that the learning process can be well described by the four components of discriminative learning as mentioned in the previous subsection. That is, a learning-to-rank method has its own input space, output space, hypothesis space, and loss function.

In the literature of machine learning, discriminative methods have been widely used to combine different kinds of features, without the necessity of defining a probabilistic framework to represent the generation of objects and the correctness of prediction. In this sense, previous works that train generative ranking models are not categorized as "learning-to-rank" methods in this book. If one has interest in such works, please refer to [45, 52, 93], etc.

Discriminative training is an automatic learning process based on the training data. This is also highly demanding for real search engines, because everyday these search engines will receive a lot of user feedback and usage logs. It is very important to automatically learn from the feedback and constantly improve the ranking mechanism.

Due to the aforementioned two characteristics, learning to rank has been widely used in commercial search engines,[18] and has also attracted great attention from the academic research community.

1.3.3 Learning-to-Rank Framework

Figure 1.6 shows the typical "learning-to-rank" flow. From the figure we can see that since learning to rank is a kind of supervised learning, a training set is needed. The creation of a training set is very similar to the creation of the test set for evaluation. For example, a typical training set consists of n training queries q_i ($i = 1, \ldots, n$), their associated documents represented by feature vectors $\mathbf{x}^{(i)} = \{x_j^{(i)}\}_{j=1}^{m^{(i)}}$ (where

---

[18]See   http://blog.searchenginewatch.com/050622-082709,   http://blogs.msdn.com/msnsearch/archive/2005/06/21/431288.aspx,   and   http://glinden.blogspot.com/2005/06/msn-search-and-learning-to-rank.html.

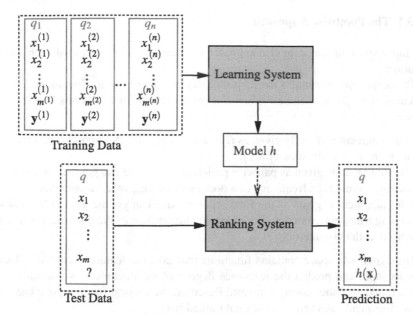

Fig. 1.6 Learning-to-rank framework

$m^{(i)}$ is the number of documents associated with query q_i), and the corresponding relevance judgments.[19] Then a specific learning algorithm is employed to learn the ranking model (i.e., the way of combining the features), such that the output of the ranking model can predict the ground truth label in the training set[20] as accurately as possible, in terms of a loss function. In the test phase, when a new query comes in, the model learned in the training phase is applied to sort the documents and return the corresponding ranked list to the user as the response to her/his query.

Many learning-to-rank algorithms can fit into the above framework. In order to better understand them, we perform a categorization on these algorithms. In particular, we group the algorithms, according to the four pillars of machine learning, into three approaches: the pointwise approach, the pairwise approach, and the listwise approach. Different approaches model the process of learning to rank in different ways. That is, they may define different input and output spaces, use different hypotheses, and employ different loss functions.

---

[19]Please distinguish the judgment for evaluation and the judgment for constructing the training set, although the process may be very similar.

[20]Hereafter, when we mention the *ground-truth labels* in the remainder of the book, we will mainly refer to the ground-truth labels in the training set, although we assume every document will have its intrinsic label, no matter whether it is judged or not.

1.3.3.1 The Pointwise Approach

The *input space* of the pointwise approach contains a feature vector of each single document.

The *output space* contains the relevance degree of each single document. Different kinds of judgments can be converted to ground truth labels in terms of relevance degree:

- If the judgment is directly given as relevance degree l_j, the ground truth label for document x_j is defined as $y_j = l_j$.
- If the judgment is given as pairwise preference $l_{u,v}$, one can get the ground truth label by counting the frequency of a document beating other documents.
- If the judgment is given as the total order π_l, one can get the ground truth label by using a mapping function. For example, the position of the document in π_l can be used as the ground truth.

The *hypothesis* space contains functions that take the feature vector of a document as input and predict the relevance degree of the document. We usually call such a function f the scoring function. Based on the scoring function, one can sort all the documents and produce the final ranked list.

The *loss function* examines the accurate prediction of the ground truth label for each single document. In different pointwise ranking algorithms, ranking is modeled as regression, classification, and ordinal regression, respectively (see Chap. 2). Therefore the corresponding regression loss, classification loss, and ordinal regression loss are used as the loss functions.

Example algorithms belonging to the pointwise approach include [13–15, 19–21, 26, 28, 44, 47, 55, 68, 78]. We will introduce some of them in Chap. 2.

Note that the pointwise approach does not consider the inter-dependency between documents, and thus the position of a document in the final ranked list is invisible to its loss function. Furthermore, the approach does not make use of the fact that some documents are actually associated with the same query. Considering that most evaluation measures for information retrieval are query level and position based, the pointwise approach has its limitations.

1.3.3.2 The Pairwise Approach

The *input space* of the pairwise approach contains pairs of documents, both represented by feature vectors.

The *output space* contains the pairwise preference (which takes values from $\{+1, -1\}$) between each pair of documents. Different kinds of judgments can be converted to ground truth labels in terms of pairwise preferences:

- If the judgment is given as relevance degree l_j, then the pairwise preference for (x_u, x_v) can be defined as $y_{u,v} = 2 \cdot I_{\{l_u \succ l_v\}} - 1$.
- If the judgment is given directly as pairwise preference, then it is straightforward to set $y_{u,v} = l_{u,v}$.

- If the judgment is given as the total order π_l, one can define $y_{u,v} = 2 \cdot I_{\{\pi_l(u) < \pi_l(v)\}} - 1$.

The *hypothesis* space contains bi-variate functions h that take a pair of documents as input and output the relative order between them. While some pairwise ranking algorithms directly define their hypotheses as such [18], in some other algorithms, the hypothesis is defined with a scoring function f for simplicity, i.e., $h(x_u, x_v) = 2 \cdot I_{\{f(x_u) > f(x_v)\}} - 1$.

The *loss function* measures the inconsistency between $h(x_u, x_v)$ and the ground truth label $y_{u,v}$. In many pairwise ranking algorithms, ranking is modeled as a pairwise classification, and the corresponding classification loss on a pair of documents is used as the loss function. When scoring function f is used, the classification loss is usually expressed in terms of the difference $(f(x_u) - f(x_v))$, rather than the thresholded quantity $h(x_u, x_v)$.

Example algorithms belonging to the pairwise approach include [8, 18, 25, 27, 37, 41, 69, 76, 94, 95]. We will introduce some of them in Chap. 3.

Note that the loss function used in the pairwise approach only considers the relative order between two documents. When one looks at only a pair of documents, however, the position of the documents in the final ranked list can hardly be derived. Furthermore, the approach ignores the fact that some pairs are generated from the documents associated with the same query. Considering that most evaluation measures for information retrieval are query level and position based, we can see a gap between this approach and ranking for information retrieval.

1.3.3.3 The Listwise Approach

The *input space* of the listwise approach contains a set of documents associated with query q, e.g., $\mathbf{x} = \{x_j\}_{j=1}^m$.

The *output space* of the listwise approach contains the ranked list (or permutation) of the documents. Different kinds of judgments can be converted to ground truth labels in terms of a ranked list:

- If the judgment is given as relevance degree l_j, then all the permutations that are consistent with the judgment are ground truth permutations. Here we define a permutation π_y as consistent with relevance degree l_j, if $\forall u, v$ satisfying $l_u > l_v$, we always have $\pi_y(u) < \pi_y(v)$. There might be multiple ground truth permutations in this case. We use Ω_y to represent the set of all such permutations.[21]
- If the judgment is given as pairwise preferences, then once again all the permutations that are consistent with the pairwise preferences are ground truth permutations. Here we define a permutation π_y as consistent with preference $l_{u,v}$, if $\forall u, v$ satisfying $l_{u,v} = +1$, we always have $\pi_y(u) < \pi_y(v)$. Again, there might also be multiple ground truth permutations in this case, and we use Ω_y to represent the set of all such permutations.

---

[21] Similar treatment can be found in the definition of Rank Correlation in Sect. 1.2.2.

- If the judgment is given as the total order π_l, one can straightforwardly define $\pi_y = \pi_l$.

Note that for the listwise approach, the output space that facilitates the learning process is exactly the same as the output space of the task. In this regard, the theoretical analysis on the listwise approach can have a more direct value to understanding the real ranking problem than the other approaches where there are mismatches between the output space that facilitates learning and the real output space of the task.

The *hypothesis* space contains multi-variate functions h that operate on a set of documents and predict their permutation. For practical reasons, the hypothesis h is usually implemented with a scoring function f, e.g., $h(\mathbf{x}) = \text{sort} \circ f(\mathbf{x})$. That is, first a scoring function f is used to give a score to each document, and then these documents are sorted in descending order of the scores to produce the desired permutation.

There are two types of *loss functions*, widely used in the listwise approach. For the first type, the loss function is explicitly related to the evaluation measures (which we call the measure-specific loss function), while for the second type, the loss function is not (which we call the non-measure-specific loss function). Note that sometimes it is not very easy to determine whether a loss function is listwise, since some building blocks of a listwise loss may also seem to be pointwise or pairwise. In this book, we mainly distinguish a listwise loss from a pointwise or pairwise loss according to the following criteria:

- A listwise loss function is defined with respect to all the training documents associated with a query.
- A listwise loss function cannot be fully decomposed to simple summation over individual documents or document pairs.
- A listwise loss function emphasizes the concept of a ranked list, and the positions of the documents in the final ranking result are visible to the loss function.

Just because of these properties of the loss function, the listwise approach is in more accordance with the ranking task in information retrieval than the pointwise and pairwise approaches.

Example algorithms that belong to the listwise approach include [10–12, 38, 59, 72, 80, 82, 84, 85, 88, 89, 96]. We will introduce some of them in Chap. 4.

To sum up, we list the key components for each approach to learning to rank in Table 1.2. In this table, for simplicity, we assume the use of a scoring function f in the hypothesis of all the approaches, although it is not always necessary to be the case.

It should be noted that although the scoring function is a kind of "pointwise" function, it is not to say that all the approaches are in nature pointwise approaches. The categorization of the aforementioned three approaches is mainly based on the four pillars of machine learning. That is, different approaches regard the same training data as in different input and output spaces, and use different loss functions and hypotheses. Accordingly, they will have difference theoretical properties. We will make more discussions on this in Part VI, with the introduction of a new theory called the *statistical learning theory for ranking*.

Table 1.2 Summary of approaches to learning to rank

Category	Pointwise		
	Regression	Classification	Ordinal regression
Input space		Single document x_j	
Output space	Real value y_j	Non-ordered category y_j	Ordered category y_j
Hypothesis space	$f(x_j)$	Classifier on $f(x_j)$	$f(x_j)$ + thresholding
Loss function	$L(f; x_j, y_j)$		

Category	Pairwise	Listwise	
	–	Non-measure-specific	Measure-specific
Input space	Document pair (x_u, x_v)	Set of documents $\mathbf{x} = \{x_j\}_{j=1}^m$	
Output space	Preference $y_{u,v}$	Ranked list π_y	
Hypothesis space	$2 \cdot I_{\{f(x_u) > f(x_v)\}} - 1$	$sort \circ f(\mathbf{x})$	
Loss function	$L(f; x_u, x_v, y_{u,v})$	$L(f; \mathbf{x}, \pi_y)$	

1.4 Book Overview

In the rest of this book, we will first give a comprehensive review on the major approaches to learning to rank in Chaps. 2, 3 and 4. For each approach, we will present the basic framework, give example algorithms, and discuss its advantages and disadvantages. It is noted that different loss functions are used in different approaches, while the same evaluation measures are used for testing their performances. Then a natural question is concerning the relationship between these loss functions and the evaluation measures. The investigation on this can help us explain the empirical results of these approaches. We will introduce such investigations in Chap. 5.

Considering that there have been some recent advances in learning to rank that cannot be simply categorized into the three major approaches, we use a separate part to introduce them. These include relational ranking, query-dependent ranking, transfer ranking, and semi-supervised ranking. Details can be found in Chaps. 6, 7, 8 and 9.

Next we introduce the benchmark datasets for the research on learning to rank in Chaps. 10 and 12, and discuss the empirical performances of typical learning-to-rank algorithms on these datasets in Chap. 11.

Then in order to fit the needs of practitioners of learning to rank, we introduce some practical issues regarding learning to rank in Chap. 13, e.g., how to mine relevance judgment from search logs, and how to select documents and features for effective training. In addition, we will also show several examples of applying learning-to-rank technologies to solve real information retrieval problems in Chap. 14.

"Nothing is more practical than theory" [77]. After introducing the algorithms and their applications, we will turn to the theoretical part of learning to rank. In particular, we will discuss the theoretical guarantee of achieving good ranking performance on unseen test data by minimizing the loss function on the training data. This is related to the generalization ability and statistical consistency of ranking methods. We will make discussions on these topics in Chaps. 15, 16, 17 and 18.

In Chaps. 19 and 20, we will summarize the book and present some future research topics.

As for the writing of the book, we do not aim to be fully rigorous. Instead we try to provide insights into the basic ideas. However, it is still unavoidable that we will use mathematics for better illustration of the problem, especially when we jump into the theoretical discussions on learning to rank. We will have to assume familiarity with basic concepts of probability theory and statistical learning in the corresponding discussions. We have listed some basics of machine learning, probability theory, algebra, and optimization in Chaps. 21 and 22. We also provide some related materials and encourage readers to refer to them in order to obtain a more comprehensive overview of the background knowledge for this book.

Throughout the book, we will use the notation rules as listed in Table 1.3. Here we would like to add one more note. Since in practice the hypothesis h is usually defined with scoring function f, we sometimes use $L(h)$ and $L(f)$ interchangeably to represent the loss function. When we need to emphasize the parameter in the scoring function f, we will use $f(w, x)$ instead of $f(x)$ in the discussion, although they actually mean the same thing. We sometimes also refer to w as the ranking model directly if there is no confusion.

1.5 Exercises

1.1 How can one estimate the size of the Web?
1.2 Investigate the relationship between the formula of BM25 and the log odds of relevance.
1.3 List different smooth functions used in LMIR, and compare them.
1.4 Use the view of the Markov process to explain the PageRank algorithm.
1.5 Enumerate all the applications of ranking that you know, in addition to document retrieval.
1.6 List the differences between generative learning and discriminative learning.
1.7 Discuss the connections between different evaluation measures for information retrieval.
1.8 Given text classification as the task, and given linear regression as the algorithms, illustrate the four components of machine learning in this case.
1.9 Discuss the major differences between ranking and classification (regression).
1.10 List the major differences between the three approaches to learning to rank.

Table 1.3 Notation rules

Meaning	Notation
Query	q or q_i
A quantity z for query q_i	$z^{(i)}$
Number of training queries	n
Number of documents associated with query q	m
Number of document pairs associated with query q	$\tilde{m}$
Feature vector of a document associated with query q	x
Feature vectors of documents associated with query q	$\mathbf{x} = \{x_j\}_{j=1}^m$
Term frequency of query q in document d	$TF(q, d)$
Inverse document frequency of query q	$IDF(q)$
Length of document d	$LEN(d)$
Hypothesis	$h(\cdot)$
Scoring function	$f(\cdot)$
Loss function	$L(\cdot)$
Expected risk	$R(\cdot)$
Empirical risk	$\hat{R}(\cdot)$
Relevance degree for document x_j	l_j
Document x_u is more relevant than document x_v	$l_u \succ l_v$
Pairwise preference between documents x_u and x_v	$l_{u,v}$
Total order of document associated with the same query	π_l
Ground-truth label for document x_j	y_j
Ground-truth label for document pair (x_u, x_v)	$y_{u,v}$
Ground-truth list for documents associate with query q	π_y
Ground-truth permutation set for documents associate with query q	Ω_y
Original document index of the jth element in permutation π	$\pi^{-1}(j)$
Rank position of document j in permutation π	$\pi(j)$
Number of classes	K
Index of class, or top positions	k
VC dimension of a function class	V
Indicator function	$I_{\{\cdot\}}$
Gain function	$G(\cdot)$
Position discount function	$\eta(\cdot)$

References

1. Amento, B., Terveen, L., Hill, W.: Does authority mean quality? Predicting expert quality ratings of web documents. In: Proceedings of the 23th Annual International ACM SIGIR Conference on Research and Development in Information Retrieval (SIGIR 2000), pp. 296–303 (2000)
2. Baeza-Yates, R., Ribeiro-Neto, B.: Modern Information Retrieval. Addison-Wesley, Reading (1999)

3. Banerjee, S., Chakrabarti, S., Ramakrishnan, G.: Learning to rank for quantity consensus queries. In: Proceedings of the 32nd Annual International ACM SIGIR Conference on Research and Development in Information Retrieval (SIGIR 2009), pp. 243–250 (2009)
4. Bartell, B., Cottrell, G.W., Belew, R.: Learning to retrieve information. In: Proceedings of Swedish Conference on Connectionism (SCC 1995) (1995)
5. Bianchini, M., Gori, M., Scarselli, F.: Inside pagerank. ACM Transactions on Internet Technologies **5**(1), 92–128 (2005)
6. Boldi, P., Santini, M., Vigna, S.: Pagerank as a function of the damping factor. In: Proceedings of the 14th International Conference on World Wide Web (WWW 2005), pp. 557–566. ACM, New York (2005)
7. Burges, C.J., Ragno, R., Le, Q.V.: Learning to rank with nonsmooth cost functions. In: Advances in Neural Information Processing Systems 19 (NIPS 2006), pp. 395–402 (2007)
8. Burges, C.J., Shaked, T., Renshaw, E., Lazier, A., Deeds, M., Hamilton, N., Hullender, G.: Learning to rank using gradient descent. In: Proceedings of the 22nd International Conference on Machine Learning (ICML 2005), pp. 89–96 (2005)
9. Cao, Y., Xu, J., Liu, T.Y., Li, H., Huang, Y., Hon, H.W.: Adapting ranking SVM to document retrieval. In: Proceedings of the 29th Annual International ACM SIGIR Conference on Research and Development in Information Retrieval (SIGIR 2006), pp. 186–193 (2006)
10. Cao, Z., Qin, T., Liu, T.Y., Tsai, M.F., Li, H.: Learning to rank: from pairwise approach to listwise approach. In: Proceedings of the 24th International Conference on Machine Learning (ICML 2007), pp. 129–136 (2007)
11. Chakrabarti, S., Khanna, R., Sawant, U., Bhattacharyya, C.: Structured learning for nonsmooth ranking losses. In: Proceedings of the 14th ACM SIGKDD International Conference on Knowledge Discovery and Data Mining (KDD 2008), pp. 88–96 (2008)
12. Chapelle, O., Wu, M.: Gradient descent optimization of smoothed information retrieval metrics. Information Retrieval Journal. Special Issue on Learning to Rank 13(3), doi:10.1007/s10791-009-9110-3 (2010)
13. Chu, W., Ghahramani, Z.: Gaussian processes for ordinal regression. Journal of Machine Learning Research **6**, 1019–1041 (2005)
14. Chu, W., Ghahramani, Z.: Preference learning with Gaussian processes. In: Proceedings of the 22nd International Conference on Machine Learning (ICML 2005), pp. 137–144 (2005)
15. Chu, W., Keerthi, S.S.: New approaches to support vector ordinal regression. In: Proceedings of the 22nd International Conference on Machine Learning (ICML 2005), pp. 145–152 (2005)
16. Ciaramita, M., Murdock, V., Plachouras, V.: Online learning from click data for sponsored search. In: Proceeding of the 17th International Conference on World Wide Web (WWW 2008), pp. 227–236 (2008)
17. Clarke, C.L., Kolla, M., Cormack, G.V., Vechtomova, O., Ashkan, A., Buttcher, S., MacKinnon, I.: Novelty and diversity in information retrieval evaluation. In: Proceedings of the 31st Annual International ACM SIGIR Conference on Research and Development in Information Retrieval (SIGIR 2008), pp. 659–666 (2008)
18. Cohen, W.W., Schapire, R.E., Singer, Y.: Learning to order things. In: Advances in Neural Information Processing Systems 10 (NIPS 1997), vol. 10, pp. 243–270 (1998)
19. Cooper, W.S., Gey, F.C., Dabney, D.P.: Probabilistic retrieval based on staged logistic regression. In: Proceedings of the 15th Annual International ACM SIGIR Conference on Research and Development in Information Retrieval (SIGIR 1992), pp. 198–210 (1992)
20. Cossock, D., Zhang, T.: Subset ranking using regression. In: Proceedings of the 19th Annual Conference on Learning Theory (COLT 2006), pp. 605–619 (2006)
21. Crammer, K., Singer, Y.: Pranking with ranking. In: Advances in Neural Information Processing Systems 14 (NIPS 2001), pp. 641–647 (2002)
22. Craswell, N., Hawking, D., Wilkinson, R., Wu, M.: Overview of the trec 2003 web track. In: Proceedings of the 12th Text Retrieval Conference (TREC 2003), pp. 78–92 (2003)
23. Deerwester, S., Dumais, S.T., Furnas, G.W., Landauer, T.K., Harshman, R.: Indexing by latent semantic analysis. Journal of the American Society for Information Science **41**, 391–407 (1990)

24. Drucker, H., Shahrary, B., Gibbon, D.C.: Support vector machines: relevance feedback and information retrieval. Information Processing and Management **38**(3), 305–323 (2002)
25. Freund, Y., Iyer, R., Schapire, R., Singer, Y.: An efficient boosting algorithm for combining preferences. Journal of Machine Learning Research **4**, 933–969 (2003)
26. Fuhr, N.: Optimum polynomial retrieval functions based on the probability ranking principle. ACM Transactions on Information Systems **7**(3), 183–204 (1989)
27. Gao, J., Qi, H., Xia, X., Nie, J.: Linear discriminant model for information retrieval. In: Proceedings of the 28th Annual International ACM SIGIR Conference on Research and Development in Information Retrieval (SIGIR 2005), pp. 290–297 (2005)
28. Gey, F.C.: Inferring probability of relevance using the method of logistic regression. In: Proceedings of the 17th Annual International ACM SIGIR Conference on Research and Development in Information Retrieval (SIGIR 1994), pp. 222–231 (1994)
29. Gyongyi, Z., Garcia-Molina, H., Pedersen, J.: Combating web spam with trustrank. In: Proceedings of the 30th International Conference on Very Large Data Bases (VLDB 2004), pp. 576–587 (2004). VLDB Endowment
30. Harrington, E.F.: Online ranking/collaborative filtering using the perceptron algorithm. In: Proceedings of the 20th International Conference on Machine Learning (ICML 2003), vol. 20(1), pp. 250–257 (2003)
31. Harrington, E.F.: Online ranking/collaborative filtering using the perceptron algorithm. In: Proceedings of the 20th International Conference on Machine Learning (ICML 2003), pp. 250–257 (2003)
32. Haveliwala, T.: Efficient computation of pageRank. Tech. rep. 1999-31, Stanford University (1999)
33. Haveliwala, T., Kamvar, S.: The second eigenvalue of the Google matrix. Tech. rep., Stanford University (2003)
34. Haveliwala, T., Kamvar, S., Jeh, G.: An analytical comparison of approaches to personalizing pagerank. Tech. rep., Stanford University (2003)
35. Haveliwala, T.H.: Topic-sensitive pagerank. In: Proceedings of the 11th International Conference on World Wide Web (WWW 2002), Honolulu, Hawaii, pp. 517–526 (2002)
36. He, B., Ounis, I.: A study of parameter tuning for term frequency normalization. In: Proceedings of the 12th International Conference on Information and Knowledge Management (CIKM 2003), pp. 10–16 (2003)
37. Herbrich, R., Obermayer, K., Graepel, T.: Large margin rank boundaries for ordinal regression. In: Advances in Large Margin Classifiers, pp. 115–132 (2000)
38. Huang, J., Frey, B.: Structured ranking learning using cumulative distribution networks. In: Advances in Neural Information Processing Systems 21 (NIPS 2008) (2009)
39. Järvelin, K., Kekäläinen, J.: IR evaluation methods for retrieving highly relevant documents. In: Proceedings of the 23rd Annual International ACM SIGIR Conference on Research and Development in Information Retrieval (SIGIR 2000), pp. 41–48 (2000)
40. Järvelin, K., Kekäläinen, J.: Cumulated gain-based evaluation of IR techniques. ACM Transactions on Information Systems **20**(4), 422–446 (2002)
41. Joachims, T.: Optimizing search engines using clickthrough data. In: Proceedings of the 8th ACM SIGKDD International Conference on Knowledge Discovery and Data Mining (KDD 2002), pp. 133–142 (2002)
42. Joachims, T.: Evaluating retrieval performance using clickthrough data. In: Text Mining, pp. 79–96 (2003)
43. Kendall, M.: Rank Correlation Methods. Oxford University Press, London (1990)
44. Kramer, S., Widmer, G., Pfahringer, B., Groeve, M.D.: Prediction of ordinal classes using regression trees. Funfamenta Informaticae **34**, 1–15 (2000)
45. Lafferty, J., Zhai, C.: Document language models, query models and risk minimization for information retrieval. In: Proceedings of the 24th Annual International ACM SIGIR Conference on Research and Development in Information Retrieval (SIGIR 2001), pp. 111–119 (2001)
46. Langville, A.N., Meyer, C.D.: Deeper inside pagerank. Internet Mathematics **1**(3), 335–400 (2004)

47. Li, P., Burges, C., Wu, Q.: McRank: Learning to rank using multiple classification and gradient boosting. In: Advances in Neural Information Processing Systems 20 (NIPS 2007), pp. 845–852 (2008)
48. Liu, T.Y., Xu, J., Qin, T., Xiong, W.Y., Li, H.: LETOR: Benchmark dataset for research on learning to rank for information retrieval. In: SIGIR 2007 Workshop on Learning to Rank for Information Retrieval (LR4IR 2007) (2007)
49. Mao, J.: Machine learning in online advertising. In: Proceedings of the 11th International Conference on Enterprise Information Systems (ICEIS 2009), p. 21 (2009)
50. Maron, M.E., Kuhns, J.L.: On relevance, probabilistic indexing and information retrieval. Journal of the ACM 7(3), 216–244 (1960)
51. McSherry, F.: A uniform approach to accelerated pagerank computation. In: Proceedings of the 14th International Conference on World Wide Web (WWW 2005), pp. 575–582. ACM, New York (2005)
52. Metzler, D.A., Croft, W.B.: A Markov random field model for term dependencies. In: Proceedings of the 28th Annual International ACM SIGIR Conference on Research and Development in Information Retrieval (SIGIR 2005), pp. 472–479 (2005)
53. Metzler, D.A., Kanungo, T.: Machine learned sentence selection strategies for query-biased summarization. In: SIGIR 2008 Workshop on Learning to Rank for Information Retrieval (LR4IR 2008) (2008)
54. Mitchell, T.: Machine Learning. McGraw-Hill, New York (1997)
55. Nallapati, R.: Discriminative models for information retrieval. In: Proceedings of the 27th Annual International ACM SIGIR Conference on Research and Development in Information Retrieval (SIGIR 2004), pp. 64–71 (2004)
56. Pavlu, V.: Large scale ir evaluation. PhD thesis, Northeastern University, College of Computer and Information Science (2008)
57. Ponte, J.M., Croft, W.B.: A language modeling approach to information retrieval. In: Proceedings of the 21st Annual International ACM SIGIR Conference on Research and Development in Information Retrieval (SIGIR 1998), pp. 275–281 (1998)
58. Qin, T., Liu, T.Y., Lai, W., Zhang, X.D., Wang, D.S., Li, H.: Ranking with multiple hyperplanes. In: Proceedings of the 30th Annual International ACM SIGIR Conference on Research and Development in Information Retrieval (SIGIR 2007), pp. 279–286 (2007)
59. Qin, T., Liu, T.Y., Li, H.: A general approximation framework for direct optimization of information retrieval measures. Information Retrieval 13(4), 375–397 (2009)
60. Qin, T., Liu, T.Y., Zhang, X.D., Chen, Z., Ma, W.Y.: A study of relevance propagation for web search. In: Proceedings of the 28th Annual International ACM SIGIR Conference on Research and Development in Information Retrieval (SIGIR 2005), pp. 408–415 (2005)
61. Qin, T., Liu, T.Y., Zhang, X.D., Wang, D., Li, H.: Learning to rank relational objects and its application to web search. In: Proceedings of the 17th International Conference on World Wide Web (WWW 2008), pp. 407–416 (2008)
62. Qin, T., Zhang, X.D., Tsai, M.F., Wang, D.S., Liu, T.Y., Li, H.: Query-level loss functions for information retrieval. Information Processing and Management 44(2), 838–855 (2008)
63. Radlinski, F., Joachims, T.: Query chain: learning to rank from implicit feedback. In: Proceedings of the 11th ACM SIGKDD International Conference on Knowledge Discovery and Data Mining (KDD 2005), pp. 239–248 (2005)
64. Richardson, M., Domingos, P.: The intelligent surfer: probabilistic combination of link and content information in pagerank. In: Advances in Neural Information Processing Systems 14 (NIPS 2001), pp. 1441–1448. MIT Press, Cambridge (2002)
65. Robertson, S.E.: Overview of the okapi projects. Journal of Documentation 53(1), 3–7 (1997)
66. Rochhio, J.J.: Relevance feedback in information retrieval. In: The SMART Retrieval System—Experiments in Automatic Document Processing, pp. 313–323 (1971)
67. Shakery, A., Zhai, C.: A probabilistic relevance propagation model for hypertext retrieval. In: Proceedings of the 15th International Conference on Information and Knowledge Management (CIKM 2006), pp. 550–558 (2006)
68. Shashua, A., Levin, A.: Ranking with large margin principles: two approaches. In: Advances in Neural Information Processing Systems 15 (NIPS 2002), pp. 937–944 (2003)

69. Shen, L., Joshi, A.K.: Ranking and reranking with perceptron. Journal of Machine Learning **60**(1–3), 73–96 (2005)
70. Singhal, A.: Modern information retrieval: a brief overview. IEEE Data Engineering Bulletin **24**(4), 35–43 (2001)
71. Surdeanu, M., Ciaramita, M., Zaragoza, H.: Learning to rank answers on large online qa collections. In: Proceedings of the 46th Annual Meeting of the Association for Computational Linguistics: Human Language Technologies (ACL-HLT 2008), pp. 719–727 (2008)
72. Talyor, M., Guiver, J., et al.: Softrank: optimising non-smooth rank metrics. In: Proceedings of the 1st International Conference on Web Search and Web Data Mining (WSDM 2008), pp. 77–86 (2008)
73. Tao, T., Zhai, C.: Regularized estimation of mixture models for robust pseudo-relevance feedback. In: Proceedings of the 29th Annual International ACM SIGIR Conference on Research and Development in Information Retrieval (SIGIR 2006), pp. 162–169 (2006)
74. Tao, T., Zhai, C.: An exploration of proximity measures in information retrieval. In: Proceedings of the 30th Annual International ACM SIGIR Conference on Research and Development in Information Retrieval (SIGIR 2007), pp. 295–302 (2007)
75. Taylor, M., Zaragoza, H., Craswell, N., Robertson, S., Burges, C.J.: Optimisation methods for ranking functions with multiple parameters. In: Proceedings of the 15th International Conference on Information and Knowledge Management (CIKM 2006), pp. 585–593 (2006)
76. Tsai, M.F., Liu, T.Y., Qin, T., Chen, H.H., Ma, W.Y.: Frank: a ranking method with fidelity loss. In: Proceedings of the 30th Annual International ACM SIGIR Conference on Research and Development in Information Retrieval (SIGIR 2007), pp. 383–390 (2007)
77. Vapnik, V.N.: Statistical Learning Theory. Wiley-Interscience, New York (1998)
78. Veloso, A., Almeida, H.M., Goçalves, M., Meira, W. Jr.: Learning to rank at query-time using association rules. In: Proceedings of the 31st Annual International ACM SIGIR Conference on Research and Development in Information Retrieval (SIGIR 2008), pp. 267–274 (2008)
79. Verberne, S., Halteren, H.V., Theijssen, D., Raaijmakers, S., Boves, L.: Learning to rank qa data. In: SIGIR 2009 Workshop on Learning to Rank for Information Retrieval (LR4IR 2009) (2009)
80. Volkovs, M.N., Zemel, R.S.: Boltzrank: learning to maximize expected ranking gain. In: Proceedings of the 26th International Conference on Machine Learning (ICML 2009), pp. 1089–1096 (2009)
81. Voorhees, E.M.: The philosophy of information retrieval evaluation. In: Lecture Notes in Computer Science (CLEF 2001), pp. 355–370 (2001)
82. Xia, F., Liu, T.Y., Wang, J., Zhang, W., Li, H.: Listwise approach to learning to rank—theorem and algorithm. In: Proceedings of the 25th International Conference on Machine Learning (ICML 2008), pp. 1192–1199 (2008)
83. Xu, J., Cao, Y., Li, H., Zhao, M.: Ranking definitions with supervised learning methods. In: Proceedings of the 14th International Conference on World Wide Web (WWW 2005), pp. 811–819. ACM Press, New York (2005)
84. Xu, J., Li, H.: Adarank: a boosting algorithm for information retrieval. In: Proceedings of the 30th Annual International ACM SIGIR Conference on Research and Development in Information Retrieval (SIGIR 2007), pp. 391–398 (2007)
85. Xu, J., Liu, T.Y., Lu, M., Li, H., Ma, W.Y.: Directly optimizing IR evaluation measures in learning to rank. In: Proceedings of the 31st Annual International ACM SIGIR Conference on Research and Development in Information Retrieval (SIGIR 2008), pp. 107–114 (2008)
86. Yang, Y.H., Hsu, W.H.: Video search reranking via online ordinal reranking. In: Proceedings of IEEE 2008 International Conference on Multimedia and Expo (ICME 2008), pp. 285–288 (2008)
87. Yang, Y.H., Wu, P.T., Lee, C.W., Lin, K.H., Hsu, W.H., Chen, H.H.: Contextseer: context search and recommendation at query time for shared consumer photos. In: Proceedings of the 16th International Conference on Multimedia (MM 2008), pp. 199–208 (2008)
88. Yeh, J.Y., Lin, J.Y., et al.: Learning to rank for information retrieval using genetic programming. In: SIGIR 2007 Workshop on Learning to Rank for Information Retrieval (LR4IR 2007) (2007)

89. Yue, Y., Finley, T., Radlinski, F., Joachims, T.: A support vector method for optimizing average precision. In: Proceedings of the 30th Annual International ACM SIGIR Conference on Research and Development in Information Retrieval (SIGIR 2007), pp. 271–278 (2007)
90. Zhai, C.: Statistical language models for information retrieval: a critical review. Foundations and Trends in Information Retrieval 2(3), 137–215 (2008)
91. Zhai, C., Cohen, W.W., Lafferty, J.: Beyond independent relevance: methods and evaluation metrics for subtopic retrieval. In: Proceedings of the 26th Annual International ACM SIGIR Conference on Research and Development in Information Retrieval (SIGIR 2003), pp. 10–17 (2003)
92. Zhai, C., Lafferty, J.: Model-based feedback in the language modeling approach to information retrieval. In: Proceedings of the 10th International Conference on Information and Knowledge Management (CIKM 2001), pp. 403–410 (2001)
93. Zhai, C., Lafferty, J.: A risk minimization framework for information retrieval. Information Processing and Management 42(1), 31–55 (2006)
94. Zheng, Z., Chen, K., Sun, G., Zha, H.: A regression framework for learning ranking functions using relative relevance judgments. In: Proceedings of the 30th Annual International ACM SIGIR Conference on Research and Development in Information Retrieval (SIGIR 2007), pp. 287–294 (2007)
95. Zheng, Z., Zha, H., Sun, G.: Query-level learning to rank using isotonic regression. In: SIGIR 2008 Workshop on Learning to Rank for Information Retrieval (LR4IR 2008) (2008)
96. Zoeter, O., Taylor, M., Snelson, E., Guiver, J., Craswell, N., Szummer, M.: A decision theoretic framework for ranking using implicit feedback. In: SIGIR 2008 Workshop on Learning to Rank for Information Retrieval (LR4IR 2008) (2008)

Part II
Major Approaches to Learning to Rank

In this part, we will introduce three major approaches to learning to rank, i.e., the pointwise, pairwise, and listwise approaches. For each approach, we will present the basic framework, give example algorithms, and discuss its advantages and disadvantages. We will then introduce the investigations of the relationship between the loss functions used in these approaches and widely used evaluation measures for information retrieval, to better understand whether it is guaranteed that these approaches can effectively optimize evaluation measures for information retrieval.

After reading this part, the readers are expected to have a big picture of the major approaches to learning to rank, and to know how to select a suitable approach/ algorithm for their given problem settings.

Chapter 2
The Pointwise Approach

Abstract In this chapter, we introduce the pointwise approach to learning to rank. Specifically, we will cover the regression-based algorithms, classification-based algorithms, and ordinal regression-based algorithms, and then make discussions on their advantages and disadvantages.

2.1 Overview

When using the technologies of machine learning to solve the problem of ranking, probably the most straightforward way is to check whether existing learning methods can be directly applied. This is exactly what the pointwise approach does. When doing so, one assumes that the exact relevance degree of each document is what we are going to predict, although this may not be necessary since the target is to produce a ranked list of the documents.

According to different machine learning technologies used, the pointwise approach can be further divided into three subcategories: regression-based algorithms, classification-based algorithms, and ordinal regression-based algorithms. For regression-based algorithms, the output space contains real-valued relevance scores; for classification-based algorithms, the output space contains non-ordered categories; and for ordinal regression-based algorithms, the output space contains ordered categories.

In the following, we will introduce representative algorithms in the three subcategories of the pointwise approach.

2.2 Regression-Based Algorithms

In this sub-category, the problem of ranking is reduced to a regression problem [5, 8]. Regression is a kind of supervised learning problem in which the target variable that one is trying to predict is continuous. When formalizing ranking as regression, one regards the relevance degree given to a document as a continuous variable, and learns the ranking function by minimizing the regret on the training set. Here we introduce a representative algorithm as an example for this sub-category.

T.-Y. Liu, *Learning to Rank for Information Retrieval*,
DOI 10.1007/978-3-642-14267-3_2, © Springer-Verlag Berlin Heidelberg 2011

Fig. 2.1 Square loss as a
function of $y_j - f(x_j)$

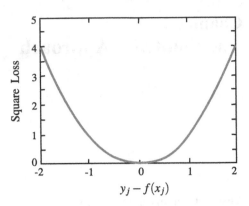

$$y_j - f(x_j)$$

2.2.1 Subset Ranking with Regression

Cossock and Zhang [5] solve the problem of ranking by reducing it to regression.

Given $\mathbf{x} = \{x_j\}_{j=1}^m$, a set of documents associated with training query q, and the ground truth labels $\mathbf{y} = \{y_j\}_{j=1}^m$ of these documents in terms of multiple ordered categories, suppose a scoring function f is used to rank these documents. The loss function is defined as the following square loss,

$$L(f; x_j, y_j) = \left(y_j - f(x_j)\right)^2. \qquad (2.1)$$

The curve of the square loss is as shown in Fig. 2.1. From the figure one can see that if and only if the output of the scoring function $f(x_j)$ exactly equals the label y_j, there is no loss. Otherwise the loss increases in a quadratic order. In other words, for a relevant document, only if the scoring function can exactly output 1, there will be zero loss. Otherwise if the output is 2 which seems to be an even stronger prediction of relevance for this document, there will be some loss. This is, in some sense, not very reasonable.

As an extension of the above method, an importance weighted regression model is further studied in [5]. The weights help the new model focus more on the regression error on the relevant documents (which are more likely to appear on top of the ranking result). Furthermore, the introduction of regularization terms to the regression loss is also investigated, which aims at making the method more generalizable (i.e., perform better on unseen test data).

In addition to the proposal of the methods, theoretical analysis on the use of the square loss function is also performed in [5]. The basic conclusion is that the square loss can upper bound the NDCG-based ranking error (see Chap. 5 for more details). However, according to the above discussions, even if there is a large regression loss, the corresponding ranking can still be optimal as long as the relative orders between the predictions $f(x_j)$ ($j = 1, \ldots, m$) are in accordance with those defined by the ground truth label. As a result, we can expect that the square loss is a loose bound of the NDCG-based ranking error.

2.3 Classification-Based Algorithms

Analogously to reducing ranking to regression, one can also consider reducing ranking to a classification problem. Classification is a kind of supervised learning problem in which the target variable that one is trying to predict is discrete. When formalizing ranking as classification, one regards the relevance degree given to a document as a category label. Here we introduce some representative algorithms in this sub-category.

2.3.1 Binary Classification for Ranking

There have been several works that study the use of a classification model for relevance ranking in information retrieval, such as [4, 9] and [16]. Here we take [16] and [9] as examples to illustrate the basic idea.

SVM-Based Method Given documents $\mathbf{x} = \{x_j\}_{j=1}^m$, and their binary relevance judgments $\mathbf{y} = \{y_j\}_{j=1}^m$ associated with a query q, one regards all the relevant documents (i.e., $y_j = +1$) as positive examples while all the irrelevant documents (i.e., $y_j = -1$) as negative examples, and therefore formulates ranking as a binary classification problem.

In particular, Support Vector Machines (SVM) [21, 22] are use to perform the classification task in [16]. The formulation of SVM is as follows when applied to the scenario of ranking (here we suppose linear scoring functions are used, i.e., $f(w, x) = w^T x$),

$$\min \frac{1}{2} \|w\|^2 + \lambda \sum_{i=1}^n \sum_{j=1}^{m^{(i)}} \xi_j^{(i)}$$

$$\text{s.t.} \quad w^T x_j^{(i)} \leq -1 + \xi_j^{(i)}, \quad \text{if } y_j^{(i)} = 0. \tag{2.2}$$

$$w^T x_j^{(i)} \geq 1 - \xi_j^{(i)}, \quad \text{if } y_j^{(i)} = 1.$$

$$\xi_j^{(i)} \geq 0, \quad j = 1, \ldots, m^{(i)}, i = 1, \ldots, n,$$

where $x_j^{(i)}, y_j^{(i)}$ are the jth document and its label with regards to query q_i and $m^{(i)}$ is the number of documents associated with query q_i.

The constraints in the above optimization problem correspond to whether each training document $x_j^{(i)}$ is classified into the right binary class. Actually the loss function behind the above optimization problem is a hinge loss defined on each document. For example, if the label $y_j^{(i)}$ is $+1$, and the model output $w^T x_j^{(i)}$ is no less than $+1$, then there is a zero loss for this document. Otherwise, the loss will be $\xi_j^{(i)}$, which is usually called the soft margin term. We plot the curve of the hinge loss in Fig. 2.2 for ease of understanding.

Fig. 2.2 Hingle loss as a
function of $y_j f(x_j)$

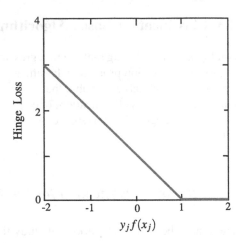

The above optimization problem can be transformed to its dual form, and get efficiently solved. In addition, the so-called kernel trick can be used to extend the algorithm to the non-linear case (for more information, please see Chaps. 21 and 22). Experiments on ad-hoc retrieval indicate that the SVM-based algorithm is comparable with and sometimes slightly better than language models. According to the experimental results, it is claimed that SVM is preferred because of its ability to learn arbitrary features automatically, fewer assumptions, and expressiveness [16].

Logistic Regression-Based Method   In [9], logistic regression (which is a popular classification technique although it contains regression in its name) is used to perform the task of binary classification for ranking. First of all, given the features of the documents, the logarithm of the odds of relevance for the document x_j is defined as follows,

$$\log\left(\frac{P(R|x_j)}{1 - P(R|x_j)}\right) = c + \sum_{t=1}^{T} w_t x_{j,t}, \tag{2.3}$$

where c is a constant.

Equivalently, the probability of relevance of a document to a query is defined as below,

$$P(R|x_j) = \frac{1}{1 + e^{-c - \sum_{t=1}^{T} w_t x_{j,t}}}. \tag{2.4}$$

Given some training data, the likelihood of the relevant documents can be computed based on the above equation, and the parameter w_t can be estimated by maximizing the likelihood. Then these parameters can be used to predict odds of relevance for other query-document pairs.

Specifically, six features are used to build the model in [9], i.e., query absolute frequency, query relative frequency, document absolute frequency, document relative frequency, relative frequency in all documents, and inverse document frequency. The derived model has been tested on typical Cranfield data collections and proven to be significantly better than a conventional vector space model.

2.3.2 Multi-class Classification for Ranking

Boosting Tree-Based Method   Li et al. [15] propose using multi-class classification to solve the problem of ranking.

Given documents $\mathbf{x} = \{x_j\}_{j=1}^m$ associated with query q, and their relevance judgment $\mathbf{y} = \{y_j\}_{j=1}^m$, suppose we have a multi-class classifier, which makes prediction $\hat{y}_j$ on x_j. Then the loss function used to learn the classifier is defined as a surrogate function of the following 0–1 classification error,

$$L(\hat{y}_j, y_j) = I_{\{y_j \neq \hat{y}_j\}}. \tag{2.5}$$

In practice, different surrogate functions of the above classification error yield different loss functions, such as the exponential loss, the hinge loss, and the logistic loss. All of them can be used to learn the classifier. In particular, in [15], the following surrogate loss function is used, and the boosting tree algorithm is employed to minimize the loss.

$$L_\phi(\hat{y}_j, y_j) = \sum_{j=1}^m \sum_{k=1}^K -\log P(\hat{y}_j = k) I_{\{y_j = k\}}. \tag{2.6}$$

Specifically, the classifier is defined with an additive model parameterized by w, i.e., $F_k(\cdot, w)$. Given a document x_j, $F_k(x_j, w)$ will indicate the degree that x_j belongs to category k. Based on F_k, the probability $P(\hat{y}_j = k)$ is defined with a logistic function,

$$P(\hat{y}_j = k) = \frac{e^{F_k(x_j, w)}}{\sum_{s=1}^K e^{F_s(x_j, w)}}. \tag{2.7}$$

In the test process, the classification results are converted into ranking scores. In particular, the output of the classifier is converted to a probability using (2.7). Suppose this probability is $P(\hat{y}_j = k)$, then the following weighted combination is used to determine the final ranking score of a document,

$$f(x_j) = \sum_{k=1}^K g(k) \cdot P(\hat{y}_j = k), \tag{2.8}$$

where $g(\cdot)$ is a monotone (increasing) function of the relevance degree k.

Association Rule Mining-Based Method   When researchers in the fields of information retrieval and machine learning investigate on different algorithms of learning to rank, data mining researchers have also made some attempts on applying well-known data mining algorithms to the problem of ranking. In particular, in [23], association rule mining is used to perform the multi-class classification for learning to rank.

The basic idea in [23] is to find those rules in the training data that can well characterize the documents belonging to the category y. Such a rule r is defined with a particular range of a feature (denoted as F, e.g., BM25 $= [0.56-0.70]$), its *support*, and *confidence*. The support is defined as the fraction of documents with relevance degree y that contain feature F, i.e., $P(y, F)$. The confidence of the rule is defined as the conditional probability of there being a document with relevance degree y given the feature F, i.e., $p(y|F)$. According to the common practice in association rule mining, only those rules whose support is larger than a threshold σ_{min} and whose confidence is also larger than a threshold θ_{min} will be finally selected. In other words, the objective of association rule mining is to minimize the following loss function:

$$-\sum_{F} H\big(P(y, F), \sigma_{min}\big) H\big(P(y|F), \theta_{min}\big), \tag{2.9}$$

where $H(a, b)$ is a step function whose value is 1 if $a \geq b$ and 0 otherwise.

Note that what we learn here is a set of rules but not a scoring function f. For the test process, it is also not as simple as applying f and sorting the documents according to their scores. Instead, one needs to proceed as follows. For a document associated with a test query, if a rule r applies to it (i.e., $P(y, F) > 0$), we will accumulate its corresponding confidence $P(y|F)$ and then perform a normalization.

$$\bar{\theta}(y) = \frac{\sum_{P(y,F)>0} P(y|F)}{\sum_{P(y,F)>0} 1}. \tag{2.10}$$

After that, the ranking score is defined as the linear combination of the normalized accumulated confidences with respect to different relevance degrees y.

$$s = \sum_{y} y \frac{\bar{\theta}(y)}{\sum_{y} \bar{\theta}(y)}. \tag{2.11}$$

Finally, all the documents are sorted according to the descending order of s.

According to the experimental results in [23], the above method has a very good ranking performance.[1]

---

[1]One should note that the ranking function used in this work is highly non-linear. This difference may partially account for the performance improvement.

2.4 Ordinal Regression-Based Algorithms

Ordinal regression takes the ordinal relationship between the ground truth labels into consideration when learning the ranking model.

Suppose there are K ordered categories. The goal of ordinal regression is to find a scoring function f, such that one can easily use thresholds $b_1 \leq b_2 \leq \cdots \leq b_{K-1}$ to distinguish the outputs of the scoring function (i.e., $f(x_j), j = 1, \ldots, m$) into different ordered categories.[2] Note that when we set $K = 2$, the ordinal regression will naturally reduce to a binary classification. In this regard, the ordinal regression-based algorithms have strong connections with classification-based algorithms.

In the literature, there are several methods in this sub-category, such as [1, 3, 6, 20], and [2]. We will introduce some of them as follows.

2.4.1 Perceptron-Based Ranking (PRanking)

PRanking is a famous algorithm on ordinal regression [6]. The goal of PRanking is to find a direction defined by a parameter vector w, after projecting the documents onto which one can easily use thresholds to distinguish the documents into different ordered categories.

This goal is achieved by means of an iterative learning process. On iteration t, the learning algorithm gets an instance x_j associated with query q. Given x_j, the algorithm predicts $\hat{y}_j = \arg\min_k \{w^T x_j - b_k < 0\}$. It then receives the ground truth label y_j. If the algorithm makes a mistake (i.e., $\hat{y}_j \neq y_j$) then there is at least one threshold, indexed by k, for which the value of $w^T x_j$ is on the wrong side of b_k. To correct the mistake, we need to move the values of $w^T x_j$ and b_k toward each other. An example is shown in Fig. 2.3. After that, the model parameter w is adjusted by $w = w + x_j$, just as in many perceptron-based algorithms. This process is repeated until the training process converges.

Harrington [10] later proposes using ensemble learning to further improve the performance of PRanking. In particular, the training data is first sub-sampled, and a PRanking model is then learned using each sample. After that, the weights and thresholds associated with the models are averaged to produce the final model. It has been proven that in this way a better generalization ability can be achieved [12].

---

[2]Note that there are some algorithms, such as [11, 13], which were also referred to as ordinal regression-based algorithms in the literature. According to our categorization, however, they belong to the pairwise approach since they do not really care about the accurate assignment of a document to one of the ordered categories. Instead, they focus more on the relative order between two documents.

Fig. 2.3 Learning process of PRanking

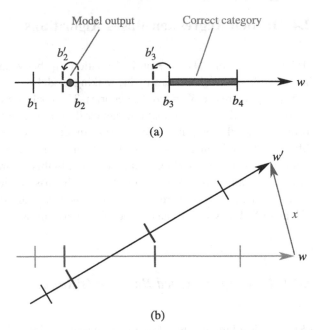

(a)

(b)

Fig. 2.4 Fixed-margin strategy

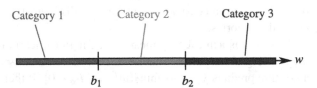

2.4.2 Ranking with Large Margin Principles

Shashua and Levin [20] try to use the SVM technology to learn the model parameter w and thresholds b_k $(k = 1, \ldots, K)$ for ordinal regression.

Specifically, two strategies are investigated. The first one is referred to as the *fixed-margin strategy*.

Given n training queries $\{q_i\}_{i=1}^n$, their associated documents $\mathbf{x}^{(i)} = \{x_j^{(i)}\}_{j=1}^{m^{(i)}}$, and the corresponding relevance judgments $\mathbf{y}^{(i)} = \{y_j^{(i)}\}_{j=1}^{m^{(i)}}$, the learning process is defined below, where the adoption of a linear scoring function is assumed. The constraints basically require every document to be correctly classified into its target ordered category, i.e., for documents in category k, $w^T x_j^{(i)}$ should exceed threshold b_{k-1} but be smaller than threshold b_k, with certain soft margins (i.e., $1 - \xi_{j,k-1}^{(i)*}$ and $1 - \xi_{j,k}^{(i)}$), respectively (see Fig. 2.4). The term $\frac{1}{2}\|w\|^2$ controls the complexity of the model w.

Fig. 2.5 Sum-of-margin strategy

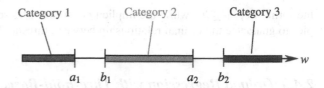

$$\min \frac{1}{2}\|w\|^2 + \lambda \sum_{i=1}^{n} \sum_{j=1}^{m^{(i)}} \sum_{k=1}^{K-1} \left(\xi_{j,k}^{(i)} + \xi_{j,k+1}^{(i)*}\right)$$

$$\begin{aligned}
\text{s.t.} \quad & w^T x_j^{(i)} - b_k \leq -1 + \xi_{j,k}^{(i)}, \quad \text{if } y_j^{(i)} = k, \\
& w^T x_j^{(i)} - b_k \geq 1 - \xi_{j,k+1}^{(i)*}, \quad \text{if } y_j^{(i)} = k+1, \qquad (2.12) \\
& \xi_{j,k}^{(i)} \geq 0, \qquad \xi_{j,k+1}^{(i)*} \geq 0, \\
& j = 1, \dots, m^{(i)}, \quad i = 1, \dots, n, \quad k = 1, \dots, K-1.
\end{aligned}$$

The second strategy is called the *sum-of-margins strategy*. In this strategy, some additional thresholds a_k are introduced, such that for category k, b_{k-1} is its lower-bound threshold and a_k is its upper-bound threshold. Accordingly, the constraints become that for documents in category k, $w^T x_j^{(i)}$ should exceed threshold b_{k-1} but be smaller than threshold a_k, with certain soft margins (i.e., $1 - \xi_{j,k-1}^{(i)*}$ and $1 - \xi_{j,k}^{(i)}$) respectively. The corresponding learning process can be expressed as follows, from which we can see that the margin term $\sum_{k=1}^{K}(a_k - b_k)$ really has the meaning of "margin" (in Fig. 2.5, $(b_k - a_k)$ is exactly the margin between category $k+1$ and category k).

$$\min \sum_{k=1}^{K-1}(a_k - b_k) + \lambda \sum_{i=1}^{n} \sum_{j=1}^{m^{(i)}} \sum_{k=1}^{K-1} \left(\xi_{j,k}^{(i)} + \xi_{j,k+1}^{(i)*}\right)$$

$$\begin{aligned}
\text{s.t.} \quad & a_k \leq b_k \leq a_{k+1}, \\
& w^T x_j^{(i)} \leq a_k + \xi_{j,k}^{(i)}, \quad \text{if } y_j^{(i)} = k, \qquad (2.13) \\
& w^T x_j^{(i)} \geq b_k - \xi_{j,k+1}^{(i)*}, \quad \text{if } y_j^{(i)} = k+1, \\
& \|w\|^2 \leq 1, \qquad \xi_{j,k}^{(i)} \geq 0, \qquad \xi_{j,k+1}^{(i)*} \geq 0, \\
& j = 1, \dots, m^{(i)}, \quad i = 1, \dots, n, \quad k = 1, \dots, K-1.
\end{aligned}$$

Ideally in the above methods, one requires b_k $(k = 1, \dots, K-1)$ to be in an increasing order (i.e., $b_{k-1} \leq b_k$). However, in practice, since there are no clear constraints on the thresholds in the optimization problem, the learning process cannot always guarantee this. To tackle the problem, Chu and Keerthi [3] propose adding explicit or implicit constraints on the thresholds. The explicit constraint simply takes the form of $b_{k-1} \leq b_k$, while the implicit constraint uses redundant training examples to guarantee the ordinal relationship between thresholds.

2.4.3 Ordinal Regression with Threshold-Based Loss Functions

In [18], different loss functions for ordinal regression are compared. Basically two types of threshold-based loss functions are investigated, i.e., immediate-threshold loss and all-threshold loss. Here, the thresholds refer to b_k ($k = 1, \ldots, K-1$) which separate different ordered categories.

Suppose the scoring function is $f(x)$, and ϕ is a margin penalty function. ϕ can be the hinge, exponential, logistic, or square function. Then the immediate-threshold loss is defined as follows:

$$L(f; x_j, y_j) = \phi\big(f(x_j) - b_{y_j-1}\big) + \phi\big(b_{y_j} - f(x_j)\big), \tag{2.14}$$

where for each labeled example (x_j, y_j), only the two thresholds defining the "correct" segment (b_{y_j-1}, b_{y_j}) are considered. In other words, the immediate-threshold loss is ignorant of whether multiple thresholds are crossed.

The all-threshold loss is defined as below, which is a sum of all threshold-violation penalties.

$$L(f; x_j, y_j) = \sum_{k=1}^{K} \phi\big(s(k, y_j)\big(b_k - f(x_j)\big)\big), \tag{2.15}$$

where

$$s(k, y_j) = \begin{cases} -1, & k < y_j, \\ +1, & k \geq y_j. \end{cases} \tag{2.16}$$

Note that the slope of the above loss function increases each time a threshold is crossed. As a result, solutions are encouraged that minimize the number of thresholds that are crossed.

The aforementioned two loss functions are tested on the MovieLens dataset.[3] The experimental results show that the all-threshold loss function can lead to a better ranking performance than multi-class classification and simple regression methods, as well as the method minimizing the immediate-threshold loss function.

2.5 Discussions

In this section, we first discuss the relationship between the pointwise approach and some early learning methods in information retrieval, such as relevance feedback. Then, we discuss the limitations of the pointwise approach.

---

[3]http://www.grouplens.org/node/73.

2.5.1 Relationship with Relevance Feedback

The pointwise approach to learning to rank, especially the classification-based algorithms, has strong correlation with the relevance feedback algorithms [7, 19]. The relevance feedback algorithms, which have played an important role in the literature of information retrieval, also leverage supervised learning technologies to improve the retrieval accuracy. The basic idea is to learn from explicit, implicit, or blind feedback to update the original query. Then the new query is used to retrieve a new set of documents. By doing this in an iterative manner, we can bring the original query closer to the optimal query so as to improve the retrieval performance.

Here we take the famous Rocchio algorithm [19] as an example to make discussions on the relationship between relevance feedback and learning to rank. The specific way that the Rocchio algorithm updates the query is as follows. First, both the query q and its associated documents are represented in a vector space. Second, through relevance feedback, $\{x_j\}_{j=1}^{m^+}$ are identified as relevant documents (i.e., $y_j = 1$), and $\{x_j\}_{j=m^++1}^{m^++m^-}$ are identified as irrelevant documents (i.e., $y_j = 0$). Third, the query vector is updated according to the following heuristic:

$$\tilde{q} = \alpha q + \beta \frac{1}{m^+} \sum_{j=1}^{m^+} x_j - \gamma \frac{1}{m^-} \sum_{j=m^++1}^{m^++m^-} x_j. \tag{2.17}$$

If we use the VSM model for retrieval, the documents will then be ranked according to their inner product with the new query vector $\tilde{q}$. Mathematically, we can define the corresponding scoring function as

$$f(x_j) = \tilde{q}^T x_j. \tag{2.18}$$

In this sense, we can regard the new query vector as the model parameter. For ease of discussion, we use w to represent this vector, i.e., $w = \tilde{q}$. Then, as shown in [14], there is actually a hidden loss function behind the above query update process, which is a function of w and x. That is,

$$L(f; x_j, y_j) = \begin{cases} \frac{1}{m^+}(\frac{1-\alpha}{4}\|w\|^2 - \beta w^T x_j), & y_j = 1, \\ \frac{1}{m^-}(\frac{1-\alpha}{4}\|w\|^2 + \gamma w^T x_j), & y_j = 0. \end{cases} \tag{2.19}$$

In other words, the Rocchio algorithm also minimizes a certain pointwise loss function. In this sense, it looks quite similar to the pointwise approach to learning to rank. However, we would like to point out its significant differences from what we call learning to rank, as shown below.

- The feature space in the Rocchio algorithm is the standard vector space as used in VSM. In this space, both query and documents are represented as vectors, and their inner product defines the relevance. In contrast, in learning to rank, the feature space contains features extracted from each query–document pair. Only documents have feature representations, and the query is not a vector in the same feature space.

- The Rocchio algorithm learns the model parameter from the feedback on a given query, and then uses the model to rank the documents associated with the same query. It does not consider the generalization of the model across queries. However, in learning to rank, we learn the ranking model from a training set, and mainly use it to rank the documents associated with unseen test queries.
- The model parameter w in the Rocchio algorithm actually has its physical meaning, i.e., it is the updated query vector. However, in learning to rank, the model parameter does not have such a meaning and only corresponds to the importance of each feature to the ranking task.
- The goal of the Rocchio algorithm is to update the query formulation for a better retrieval but not to learn an optimal ranking function. In other words, after the query is updated, the fixed ranking function (e.g., the VSM model) is used to return a new set of related documents.

2.5.2 Problems with the Pointwise Approach

Since the input object in the pointwise approach is a single document, the relative order between documents cannot be naturally considered in their learning processes. However ranking is more about predicting relative order than accurate relevance degree.

Furthermore, the two intrinsic properties of the evaluation measures for ranking (i.e., query level and position based) cannot be well considered by the pointwise approach:

1. The fact is ignored in these algorithms that some documents are associated with the same query and some others are not. As a result, when the number of associated documents varies largely for different queries,[4] the overall loss function will be dominated by those queries with a large number of documents.
2. The position of each document in the ranked list is invisible to the pointwise loss functions. Therefore, the pointwise loss function may unconsciously emphasize too much those unimportant documents (which are ranked low in the final ranked list and thus do not affect user experiences).

2.5.3 Improved Algorithms

In order to avoid the problems with the pointwise approach as mentioned above, RankCosine [17] introduces a query-level normalization factor to the pointwise loss

---

[4]For the re-ranking scenario, the number of documents to rank for each query may be very similar, e.g., the top 1000 documents per query. However, if we consider all the documents containing the query word, the difference between the number of documents for popular queries and that for tail queries may be very large.

function. In particular, it defines the loss function based on the cosine similarity between the score vector outputted by the scoring function f for query q, and the score vector defined with the ground truth label (referred to as the *cosine loss* for short). That is,

$$L(f; \mathbf{x}, \mathbf{y}) = \frac{1}{2} - \frac{\sum_{j=1}^{m} \varphi(y_j)\varphi(f(x_j))}{2\sqrt{\sum_{j=1}^{m} \varphi^2(y_j)}\sqrt{\sum_{j=1}^{m} \varphi^2(f(x_j))}} \tag{2.20}$$

where φ is a transformation function, which can be linear, exponential, or logistic.

After defining the cosine loss, the gradient descent method is used to perform the optimization and learn the scoring function.

According to [17], the so-defined cosine loss has the following properties.

- The cosine loss can be regarded as a kind of regression loss, since it requires the prediction on the relevance of a document to be as close to the ground truth label as possible.
- Because of the query-level normalization factor (the denominator in the loss function), the cosine loss is insensitive to the varying numbers of documents with respect to different queries.
- The cosine loss is bounded between 0 and 1, thus the overall loss on the training set will not be dominated by specific hard queries.
- The cosine loss is scale invariant. That is, if we multiply all the ranking scores outputted by the scoring function by the same constant, the cosine loss will not change. This is quite in accordance with our intuition on ranking.

2.6 Summary

In this chapter, we have introduced various pointwise ranking methods, and discussed their relationship with previous learning-based information retrieval models, and their limitations.

So far, the pointwise approach can only be a sub-optimal solution to ranking. To tackle the problem, researchers have made attempts on regarding document pairs or the entire set of documents associated with the same query as the input object. This results in the pairwise and listwise approaches to learning to rank. With the pairwise approach, the relative order among documents can be better modeled. With the listwise approach, the positional information can be visible to the learning-to-rank process.

2.7 Exercises

2.1 Enumerate widely used loss functions for classification, and prove whether they are convex.

2.2 A regularization item is usually introduced when performing regression. Show the contribution of this item to the learning process.

2.3 Please list the major differences between ordinal regression and classification.

2.4 What kind of information is missing after reducing ranking to ordinal regression, classification, or regression?

2.5 Survey more algorithms for ordinal regression in addition to those introduced in this chapter.

2.6 What is the inherent loss function in the algorithm PRanking?

References

1. Chu, W., Ghahramani, Z.: Gaussian processes for ordinal regression. Journal of Machine Learning Research **6**, 1019–1041 (2005)
2. Chu, W., Ghahramani, Z.: Preference learning with Gaussian processes. In: Proceedings of the 22nd International Conference on Machine Learning (ICML 2005), pp. 137–144 (2005)
3. Chu, W., Keerthi, S.S.: New approaches to support vector ordinal regression. In: Proceedings of the 22nd International Conference on Machine Learning (ICML 2005), pp. 145–152 (2005)
4. Cooper, W.S., Gey, F.C., Dabney, D.P.: Probabilistic retrieval based on staged logistic regression. In: Proceedings of the 15th Annual International ACM SIGIR Conference on Research and Development in Information Retrieval (SIGIR 1992), pp. 198–210 (1992)
5. Cossock, D., Zhang, T.: Subset ranking using regression. In: Proceedings of the 19th Annual Conference on Learning Theory (COLT 2006), pp. 605–619 (2006)
6. Crammer, K., Singer, Y.: Pranking with ranking. In: Advances in Neural Information Processing Systems 14 (NIPS 2001), pp. 641–647 (2002)
7. Drucker, H., Shahrary, B., Gibbon, D.C.: Support vector machines: relevance feedback and information retrieval. Information Processing and Management **38**(3), 305–323 (2002)
8. Fuhr, N.: Optimum polynomial retrieval functions based on the probability ranking principle. ACM Transactions on Information Systems **7**(3), 183–204 (1989)
9. Gey, F.C.: Inferring probability of relevance using the method of logistic regression. In: Proceedings of the 17th Annual International ACM SIGIR Conference on Research and Development in Information Retrieval (SIGIR 1994), pp. 222–231 (1994)
10. Harrington, E.F.: Online ranking/collaborative filtering using the perceptron algorithm. In: Proceedings of the 20th International Conference on Machine Learning (ICML 2003), pp. 250–257 (2003)
11. Herbrich, R., Obermayer, K., Graepel, T.: Large margin rank boundaries for ordinal regression. In: Advances in Large Margin Classifiers, pp. 115–132 (2000)
12. Herbrich, R., Graepel, T., Campbell, C.: Bayes point machines. Journal of Machine Learning Research **1**, 245–279 (2001)
13. Joachims, T.: Optimizing search engines using clickthrough data. In: Proceedings of the 8th ACM SIGKDD International Conference on Knowledge Discovery and Data Mining (KDD 2002), pp. 133–142 (2002)
14. Li, F., Yang, Y.: A loss function analysis for classification methods in text categorization. In: Proceedings of the 20th International Conference on Machine Learning (ICML 2003), pp. 472–479 (2003)
15. Li, P., Burges, C., Wu, Q.: McRank: learning to rank using multiple classification and gradient boosting. In: Advances in Neural Information Processing Systems 20 (NIPS 2007), pp. 845–852 (2008)
16. Nallapati, R.: Discriminative models for information retrieval. In: Proceedings of the 27th Annual International ACM SIGIR Conference on Research and Development in Information Retrieval (SIGIR 2004), pp. 64–71 (2004)

17. Qin, T., Zhang, X.D., Tsai, M.F., Wang, D.S., Liu, T.Y., Li, H.: Query-level loss functions for information retrieval. Information Processing and Management **44**(2), 838–855 (2008)
18. Rennie, J.D.M., Srebro, N.: Loss functions for preference levels: regression with discrete ordered labels. In: IJCAI 2005 Multidisciplinary Workshop on Advances in Preference Handling. ACM, New York (2005)
19. Rochhio, J.J.: Relevance feedback in information retrieval. In: The SMART Retrieval System—Experiments in Automatic Document Processing, pp. 313–323 (1971)
20. Shashua, A., Levin, A.: Ranking with large margin principles: two approaches. In: Advances in Neural Information Processing Systems 15 (NIPS 2002), pp. 937–944 (2003)
21. Vapnik, V.N.: The Nature of Statistical Learning Theory. Springer, Berlin (1995)
22. Vapnik, V.N.: Statistical Learning Theory. Wiley-Interscience, New York (1998)
23. Veloso, A., de Almeida, H.M., Goncalves, M.A., Wagner, M. Jr.: Learning to rank at query-time using association rules. In: Proceedings of the 31st Annual International ACM SIGIR Conference on Research and Development in Information Retrieval (SIGIR 2008), pp. 267–274 (2008)

Chapter 3
The Pairwise Approach

Abstract In this chapter we will introduce the pairwise approach to learning to rank. Specifically we first introduce several example algorithms, whose major differences are in the loss functions. Then we discuss the limitations of these algorithms and present some improvements that enable better ranking performance.

3.1 Overview

The pairwise approach does not focus on accurately predicting the relevance degree of each document; instead, it cares about the relative order between two documents. In this sense, it is closer to the concept of "ranking" than the pointwise approach.

In the pairwise approach, ranking is usually reduced to a classification on document pairs, i.e., to determine which document in a pair is preferred. That is, the goal of learning is to minimize the number of miss-classified document pairs. In the extreme case, if all the document pairs are correctly classified, all the documents will be correctly ranked. Note that this classification differs from the classification in the pointwise approach, since it operates on every two documents under investigation. A natural concern is that document pairs are not independent, which violates the basic assumption of classification. The fact is that although in some cases this assumption really does not hold, one can still use classification technology to learn the ranking model. However, a different theoretical framework is needed to analyze the generalization of the learning process. We will make discussions on this in Part VI.

There are many pairwise ranking algorithms proposed in the literature of learning to rank, based on neural network [8], perceptron [21, 30], Boosting [18], support vector machines [22, 23], and other learning machines [12, 33, 36, 37]. In the rest of this chapter, we will introduce several examples of them.

3.2 Example Algorithms

3.2.1 Ordering with Preference Function

In [12], a hypothesis $h(x_u, x_v)$ directly defined on a pair of documents is studied (i.e., without use of the scoring function f). In particular, given two documents x_u

Fig. 3.1 Loss function
in [12]

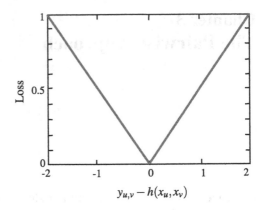

and x_v associated with a training query q, the loss function is defined below,

$$L(h; x_u, x_v, y_{u,v}) = \frac{|y_{u,v} - h(x_u, x_v)|}{2}, \tag{3.1}$$

where the hypothesis is defined as $h(x_u, x_v) = \sum_t w_t h_t(x_u, x_v)$ and $h_t(x_u, x_v)$ is called the base preference function.

Suppose $h(x_u, x_v)$ takes a value from $[-1, +1]$, where a positive value indicates that document x_u is ranked before x_v, and a negative value indicates the opposite. According to the above loss function, we can easily find that if the ground truth label indicates that document x_u should be ranked before document x_v (i.e., $y_{u,v} = 1$) but $h(x_u, x_v) \neq 1$, there will be a loss for this pair of documents. Similarly, if the ground truth label indicates that document x_u should be ranked after document x_v (i.e., $y_{u,v} = -1$) but $h(x_u, x_v) \neq -1$, there will also be a loss. The curve of the above loss function is shown in Fig. 3.1.

With the above loss function, the weighted majority algorithm, e.g., the Hedge algorithm, is used to learn the parameters in the hypothesis h. Note that the hypothesis h is actually a preference function, which cannot directly output the ranked list of the documents. In this case, an additional step is needed to convert the pairwise preference between any two documents to the total order of these documents. To this end, one needs to find the ranked list σ, which has the largest agreement with the pairwise preferences. This process is described by

$$\sigma = \max_\pi \sum_{u<v} h(x_{\pi^{-1}(u)}, x_{\pi^{-1}(v)}). \tag{3.2}$$

As we know, this is a typical problem, called rank aggregation. It has been proven NP-hard to find the optimal solution to the above optimization problem. To tackle the challenge, a greedy ordering algorithm is used in [12], which can be much more efficient, and its agreement with the pairwise preferences is at least half the agreement of the optimal algorithm.

One may have noticed a problem in the above formulation. When defining the loss function for learning, attention is paid to the preference function. However,

Fig. 3.2 The network
structure for the preference
function

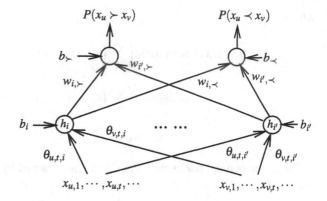

the true ranker used in the final test is defined by (3.2). It is not obvious whether
the optimal preference function learned by loss minimization can really lead to the
optimal ranker. To answer this concern, in [12], it has been proven that the loss
with respect to the ranker can be bounded by the loss with respect to the preference
function, as follows:

$$L(\sigma, y) \leq \frac{\sum_{u<v}(1 - h(x_{\sigma^{-1}(u)}, x_{\sigma^{-1}(v)})) + \sum_{u,v:y_{u,v}=1} L(h; x_u, x_v, y_{u,v})}{\sum_{u,v:y_{u,v}=1} 1} \tag{3.3}$$

3.2.2 SortNet: Neural Network-Based Sorting Algorithm

The goal of [28] is also to learn a preference function from training data. Specifically, a neural network with the structure as shown in Fig. 3.2 is used as the preference function.

As can be seen from Fig. 3.2, there are two outputs of the network, $P(x_u \succ x_v)$ and $P(x_u \prec x_v)$. Since the network approximates a preference function, the
following constraints on these two outputs are enforced:

$$P(x_u \succ x_v) = P(x_v \prec x_u). \tag{3.4}$$

Specifically, these two outputs are generated as follows:

$$P(x_u \succ x_v) = \text{sigmoid}\left(\sum_{i,i'}(w_{i,\succ}h_i(x_u, x_v) + w_{i',\succ}h_{i'}(x_u, x_v) + b_\succ)\right)$$

$$= \text{sigmoid}\left(\sum_{i,i'}(w_{i,\prec}h_i(x_v, x_u) + w_{i',\prec}h_{i'}(x_v, x_u) + b_\prec)\right)$$

$$= P(x_v \prec x_u), \tag{3.5}$$

where

$$h_i(x_u, x_v) = \text{sigmoid}\left(\sum_t \left(\theta_{u,t,i} x_{u,t} + \theta_{v,t,i} x_{v,t} + b_i\right)\right)$$

$$= \text{sigmoid}\left(\sum_t \left(\theta_{u,t,i'} x_{v,t} + \theta_{v,t,i'} x_{u,t} + b_{i'}\right)\right)$$

$$= h_{i'}(x_v, x_u). \tag{3.6}$$

Then, the optimal parameters θ, w, and b are learned by minimizing the following loss function:

$$L(h; x_u, x_v, y_{u,v}) = \left(y_{u,v} - P(x_u \succ x_v)\right)^2 + \left(y_{v,u} - P(x_u \prec x_v)\right)^2. \tag{3.7}$$

For testing, the learned preference function is used to generate pairwise preferences for all possible document pairs. Then an additional sorting (aggregation) step, just as in [12], is used to resolve the conflicts in these pairwise preferences and to generate a final ranked list.

3.2.3 RankNet: Learning to Rank with Gradient Descent

RankNet [8] is one of the learning-to-rank algorithms used by commercial search engines.[1]

In RankNet, the loss function is also defined on a pair of documents, but the hypothesis is defined with the use of a scoring function f. Given two documents x_u and x_v associated with a training query q, a target probability $\bar{P}_{u,v}$ is constructed based on their ground truth labels. For example, we can define $\bar{P}_{u,v} = 1$, if $y_{u,v} = 1$; $\bar{P}_{u,v} = 0$, otherwise. Then, the modeled probability $P_{u,v}$ is defined based on the difference between the scores of these two documents given by the scoring function, i.e.,

$$P_{u,v}(f) = \frac{\exp(f(x_u) - f(x_v))}{1 + \exp(f(x_u) - f(x_v))}. \tag{3.8}$$

Then the cross entropy between the target probability and the modeled probability is used as the loss function, which we refer to as the *cross entropy loss* for short.

$$L(f; x_u, x_v, y_{u,v}) = -\bar{P}_{u,v} \log P_{u,v}(f) - (1 - \bar{P}_{u,v}) \log\left(1 - P_{u,v}(f)\right). \tag{3.9}$$

---

[1] As far as we know, Microsoft Bing Search (http://www.bing.com/) is using the model trained with a variation of RankNet.

Fig. 3.3 A two-layer neural network

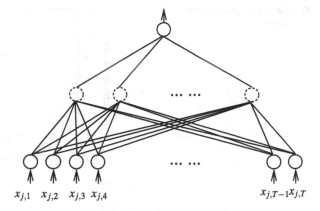

$$x_{j,1} \quad x_{j,2} \quad x_{j,3} \quad x_{j,4} \qquad\qquad x_{j,T-1} x_{j,T}$$

It is not difficult to verify that the cross entropy loss is an upper bound of the pairwise 0–1 loss, which is defined by

$$L_{0-1}(f; x_u, x_v, y_{u,v}) = \begin{cases} 1, & y_{u,v}(f(x_u) - f(x_v)) < 0, \\ 0, & \text{otherwise.} \end{cases} \tag{3.10}$$

A neural network is then used as the model and gradient descent as the optimization algorithm to learn the scoring function f. A typical two-layer neural network is shown in Fig. 3.3, where the features of a document are inputted at the bottom layer; the second layer consists of several hidden nodes; each node involves a sigmoid transformation; and the output of the network is the ranking score of the document.

In [25], a nested ranker is built on top of RankNet to further improve the retrieval performance. Specifically, the new method iteratively re-ranks the top scoring documents. At each iteration, this approach uses the RankNet algorithm to re-rank a subset of the results. This splits the problem into smaller and easier tasks and generates a new distribution of the results to be learned by the algorithm. Experimental results show that making the learning algorithm iteratively concentrate on the top scoring results can improve the accuracy of the top ten documents.

3.2.4 FRank: Ranking with a Fidelity Loss

Some problems with the loss function used in RankNet have been pointed out in [33]. Specifically, the curve of the cross entropy loss as a function of $f(x_u) - f(x_v)$ is plotted in Fig. 3.4. From this figure, one can see that in some cases the cross entropy loss has a non-zero minimum, indicating that there will always be some loss no matter what kind of model is used. This may not be in accordance with our intuition of a loss function. Furthermore, the loss is not bounded, which may lead to the dominance of some difficult document pairs in the training process.

Fig. 3.4 Cross entropy loss
as a function of
$f(x_u) - f(x_v)$

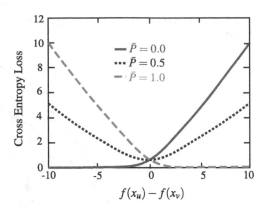

To tackle these problems, a new loss function named the *fidelity loss* is proposed
in [33], which has the following form:

$$L(f; x_u, x_v, y_{u,v}) = 1 - \sqrt{\bar{P}_{u,v} P_{u,v}(f)} - \sqrt{(1 - \bar{P}_{u,v})(1 - P_{u,v}(f))}. \quad (3.11)$$

The fidelity was originally used in quantum physics to measure the difference
between two probabilistic states of a quantum. When being used to measure the difference between the target probability and the modeled probability, the fidelity loss
has the shape as shown in Fig. 3.5 as a function of $f(x_u) - f(x_v)$. By comparing the
fidelity loss with the cross entropy loss, we can see that the fidelity loss is bounded
between 0 and 1, and always has a zero minimum. These properties are nicer than
those of the cross entropy loss. On the other hand, however, while the cross entropy
loss is convex, the fidelity loss becomes non-convex. Such a non-convex objective
is more difficult to optimize and one needs to be careful when performing the optimization. Furthermore, the fidelity loss is no longer an upper bound of the pairwise
0–1 loss.

In [33], a generalized additive model is proposed as the ranking function, and a
technique similar to Boosting is used to learn the coefficients in the additive model.
In particular, in each iteration, a new weak ranker (e.g., a new feature) is added,
and the combination coefficient is set by considering the gradient of the fidelity
loss with respect to it. When the addition of a new ranker does not bring in significant reduction of the loss any more, the learning process converges. According to
the experimental results reported in [33], FRank outperforms RankNet on several
datasets.

3.2.5 RankBoost

The method of RankBoost [18] adopts AdaBoost [19] for the classification over
document pairs. The only difference between RankBoost and AdaBoost is that the
distribution in RankBoost is defined on document pairs while that in AdaBoost is
defined on individual documents.

Fig. 3.5 Fidelity loss as a function of $f(x_u) - f(x_v)$

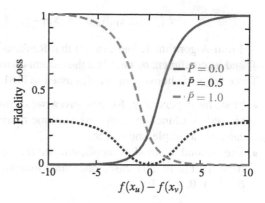

Algorithm 1: Learning algorithm for RankBoost

Input: training data in terms of document pairs
Given: initial distribution $\mathcal{D}_1$ on input document pairs.
For $t = 1, \ldots, T$
    Train weak ranker f_t based on distribution $\mathcal{D}_t$.
    Choose α_t
    Update $\mathcal{D}_{t+1}(x_u^{(i)}, x_v^{(i)}) = \frac{1}{Z_t} \mathcal{D}_t(x_u^{(i)}, x_v^{(i)}) \exp(\alpha_t(f_t(x_u^{(i)}) - f_t(x_v^{(i)})))$
    where $Z_t = \sum_{i=1}^n \sum_{u,v:y_{u,v}^{(i)}=1} \mathcal{D}_t(x_u^{(i)}, x_v^{(i)}) \exp(\alpha_t(f_t(x_u^{(i)}) - f_t(x_v^{(i)})))$.
Output: $f(x) = \sum_t \alpha_t f_t(x)$.

Fig. 3.6 Exponential loss for RankBoost

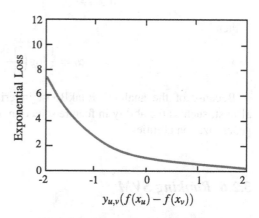

The algorithm flow of RankBoost is given in Algorithm 1, where $\mathcal{D}_t$ is the distribution on document pairs, f_t is the weak ranker selected at the tth iteration, and α_t is the weight for linearly combining the weak rankers. The RankBoost algorithm actually minimizes the exponential loss defined below (also shown in Fig. 3.6). It is clear that this exponential loss is also an upper bound of the pairwise 0–1 loss.

$$L(f; x_u, x_v, y_{u,v}) = \exp(-y_{u,v}(f(x_u) - f(x_v))). \tag{3.12}$$

From Algorithm 1, one can see that RankBoost learns the optimal weak ranker f_t and its coefficient α_t based on the current distribution of the document pairs ($\mathcal{D}_t$). Three ways of choosing α_t are discussed in [18].

- First, most generally, for any given weak ranker f_t, it can be shown that Z_t, viewed as a function of α_t, has a unique minimum, which can be found numerically via a simple line search.
- The second method is applicable in the special case that f_t takes a value from $\{0,1\}$. In this case, one can minimize Z_t analytically as follows. For $b \in \{-1, 0, +1\}$, let

$$W_{t,b} = \sum_{i=1}^{n} \sum_{u,v:y_{u,v}^{(i)}=1} \mathcal{D}_t\left(x_u^{(i)}, x_v^{(i)}\right) I_{\{f_t(x_u^{(i)}) - f_t(x_v^{(i)})=b\}}. \tag{3.13}$$

Then

$$\alpha_t = \frac{1}{2} \log\left(\frac{W_{t,-1}}{W_{t,+1}}\right). \tag{3.14}$$

- The third way is based on the approximation of Z_t, which is applicable when f_t takes a real value from $[0, 1]$. In this case, if we define

$$r_t = \sum_{i=1}^{n} \sum_{u,v:y_{u,v}^{(i)}=1} \mathcal{D}_t\left(x_u^{(i)}, x_v^{(i)}\right)\left(f_t\left(x_u^{(i)}\right) - f_t\left(x_v^{(i)}\right)\right), \tag{3.15}$$

then

$$\alpha_t = \frac{1}{2} \log\left(\frac{1+r_t}{1-r_t}\right). \tag{3.16}$$

Because of the analogy, RankBoost inherits many nice properties from AdaBoost, such as the ability in feature selection, convergence in training, and certain generalization abilities.

3.2.6 Ranking SVM

Ranking SVM [22, 23] applies the SVM technology to perform pairwise classification.[2] Given n training queries $\{q_i\}_{i=1}^{n}$, their associated document pairs $(x_u^{(i)}, x_v^{(i)})$,

---

[2]Note that Ranking SVM was originally proposed in [22] to solve the problem of ordinal regression. However, according to its formulation, it solves the problem of pairwise classification in an even more natural way.

Fig. 3.7 Hinge loss for
Ranking SVM

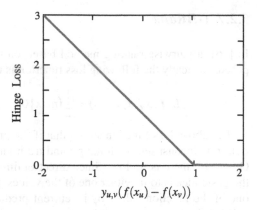

and the corresponding ground truth label $y_{u,v}^{(i)}$, the mathematical formulation of
Ranking SVM is as shown below, where a linear scoring function is used, i.e.,
$f(x) = w^T x$,

$$\min \frac{1}{2} \|w\|^2 + \lambda \sum_{i=1}^{n} \sum_{u,v:y_{u,v}^{(i)}=1} \xi_{u,v}^{(i)}$$

$$\text{s.t.} \quad w^T \left(x_u^{(i)} - x_v^{(i)} \right) \geq 1 - \xi_{u,v}^{(i)}, \text{if } y_{u,v}^{(i)} = 1, \qquad (3.17)$$

$$\xi_{u,v}^{(i)} \geq 0, \quad i = 1, \ldots, n.$$

As we can see, the objective function in Ranking SVM is very similar to that in
SVM, where the term $\frac{1}{2}\|w\|^2$ controls the complexity of the model w. The differ-
ence between Ranking SVM and SVM lies in the constraints, which are constructed
from document pairs. The loss function in Ranking SVM is a hinge loss defined on
document pairs. For example, for a training query q, if document x_u is labeled as
being more relevant than document x_v (mathematically, $y_{u,v} = 1$), then if $w^T x_u$ is
larger than $w^T x_v$ by a margin of 1, there is no loss. Otherwise, the loss will be $\xi_{u,v}$.
Such a hinge loss is an upper bound of the pairwise 0–1 loss. The curve of the hinge
loss is plotted in Fig. 3.7.

Since Ranking SVM is well rooted in the framework of SVM, it inherits nice
properties of SVM. For example, with the help of margin maximization, Ranking SVM can have a good generalization ability. Kernel tricks can also be applied
to Ranking SVM, so as to handle complex non-linear problems. Furthermore, recently, several fast implementations of Ranking SVM have also been developed,
such as [11] and [24]. The corresponding tools have also been available online.[3]

---

[3] See http://olivier.chapelle.cc/primal/ and http://www.cs.cornell.edu/People/tj/svm_light/svm_
rank.html.

3.2.7 GBRank

In [36], a pairwise ranking method based on Gradient Boosting Tree [20] is proposed. Basically the following loss function is used,

$$L(f; x_u, x_v, y_{u,v}) = \frac{1}{2}\left(\max\{0, y_{u,v}(f(x_v) - f(x_u))\}\right)^2. \tag{3.18}$$

The above loss function shows that if the predicted preference by scoring function f is inconsistent with the ground truth label $y_{u,v}$, there will be a square loss. Otherwise there is no loss. Since direct optimization of the above loss is difficult, the basic idea is to fix either one of the values $f(x_u)$ and $f(x_v)$, e.g., replace either one of the function values by its current predicted value, and solve the remaining problem by means of regression.

To avoid obtaining an optimal scoring function which is constant, a regularization item is further introduced as follows,

$$L(f; x_u, x_v, y_{u,v}) = \frac{1}{2}\left(\max\{0, y_{u,v}(f(x_v) - f(x_u)) + \tau\}\right)^2 - \lambda\tau^2. \tag{3.19}$$

To summarize, Algorithm 2 named GBRank is used to perform the ranking function learning.

Algorithm 2: Learning algorithm for GBRank

Input: training data in terms of document pairs
Given: initial guess of the ranking function f_0.
For $t = 1, \ldots, T$
    using f_{t-1} as the current approximation of f, we separate the training data
    into two disjoint sets.
    $S^+ = \{\langle x_u, x_v\rangle | f_{t-1}(x_u) \geq f_{t-1}(x_v) + \tau\}$
    $S^- = \{\langle x_u, x_v\rangle | f_{t-1}(x_u) < f_{t-1}(x_v) + \tau\}$
    fitting a regression function $g_t(x)$ using gradient boosting tree [20] and the
    following training data
    $\{(x_u, f_{t-1}(x_v) + \tau), (x_v, f_{t-1}(x_u) - \tau) | \langle x_u, x_v\rangle \in S^-\}$
    forming (with normalization of the range of f_t)
    $f_t(x) = \frac{t \cdot f_{t-1}(x) + \eta g_t(x)}{t+1}$
    where η is a shrinkage factor.
Output: $f(x) = f_T(x)$.

In [38], some improvements over GBRank are made. Specifically, quadratic approximation to the loss function in GBRank is conducted based on Taylor expansion, and then an upper bound of the loss function, which is easier to optimize, is derived. Experimental results show that the new method can outperform GBRank in many settings.

In [39], the idea of GBRank is further extended, and those pairs of documents with the same label (i.e., the so-called ties) are also considered in the loss function. For this purpose, well-known pairwise comparison models are used to define the loss function, such as the Bradley-Terry model and the Thurstone–Mosteller model [6, 32]. Again, a gradient boosting tree is used as the algorithm to minimize the loss function and learn the ranking function.

3.3 Improved Algorithms

It seems that the pairwise approach has its advantages as compared to the pointwise approach, since it models the relative order between documents rather than the absolute value of the relevance degree of a document. However, the pairwise approach also has its disadvantages.

1. When we are given the relevance judgment in terms of multiple ordered categories, however, converting it to pairwise preference will lead to the loss of information about the finer granularity in the relevance judgment. Since any two documents with different relevance degrees can construct a document pair, we may have such a document pair whose original relevance degrees are *Excellent* and *Bad*, and another document pair whose original relevance degrees are *Fair* and *Bad*. It is clear that these two pairs represent different magnitudes of preferences; however, in the algorithms introduced above, they are treated equally without distinction.
2. Since the number of pairs can be as large as in the quadratic order of the number of documents, the imbalanced distribution across queries may be even more serious for the pairwise approach than the pointwise approach.
3. The pairwise approach is more sensitive to noisy labels than the pointwise approach. That is, a noisy relevance judgment on a single document can lead to a large number of mis-labeled document pairs.
4. Most of the pairwise ranking algorithms introduced above do not consider the position of documents in the final ranking results, but instead define their loss functions directly on the basis of individual document pairs.

These problems may affect the effectiveness of the pairwise approach. To tackle the problems, several attempts have been proposed, which will be depicted as follows.

3.3.1 Multiple Hyperplane Ranker

To tackle the first problem of the pairwise approach as aforementioned, Qin et al. [26] propose using a new algorithm named Multiple Hyperplane Ranker (MHR). The basic idea is "divide-and-conquer". Suppose there are K different categories of judgments, then one can train $K(K - 1)/2$ Ranking SVM models in total, with each model trained from the document pairs with two categories of judgments (see Fig. 3.8). At the test phase, rank aggregation is used to merge the ranking results given by each model to produce the final ranking result. For instance, suppose that

Fig. 3.8 Training multiple rankers

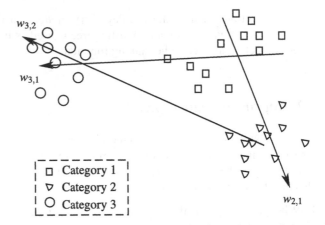

the model trained from categories k and l is denoted by $f_{k,l}$, then the final ranking results can be attained by using the weighted BordaCount aggregation.[4]

$$f(x) = \sum_{k,l} \alpha_{k,l} f_{k,l}(x). \tag{3.20}$$

Here the combination coefficient $\alpha_{k,l}$ can be pre-specified or learned from a separate validation set. The experimental results in [26] show that by considering more information about the judgment, the ranking performance can be significantly improved over Ranking SVM. Note that the technology used in MHR actually can be extended to any other pairwise ranking algorithms.

3.3.2 Magnitude-Preserving Ranking

In [14], Cortes et al. also attempt to tackle the first problem with the pairwise approach. In particular, they propose keeping the magnitude of the labeled preferences, and accordingly use the so-called *magnitude-preserving loss* (MP loss), *hinge magnitude-preserving loss* (HMP loss), and SVM regression loss (SVR loss) for learning to rank. These three loss functions can effectively penalize a pairwise misranking by the magnitude of predicted preference or the βth power of that magnitude. That is, considering the property of the power function x^β, if the loss on a pair is large, it will be highly penalized if β is set to be large.

---

[4]Note that there are many algorithms for rank aggregation proposed in the literature, such as BordaCount [2, 5, 16], median rank aggregation [17], genetic algorithm [4], fuzzy logic-based rank aggregation [1], and Markov chain-based rank aggregation [16]. Although BordaCount is used in [26] as an example, it by no means dictates that other methods cannot be used for the same purpose.

The MP loss is defined as follows:

$$L_{MP}(f; x_u, x_v, y_{u,v}) = \left| (f(x_u) - f(x_v)) - y_{u,v} \right|^\beta. \tag{3.21}$$

The HMP loss is defined as follows:

$$L_{HMP}(f; x_u, x_v, y_{u,v}) = \begin{cases} 0, & y_{u,v}(f(x_u) - f(x_v)) \ge 0, \\ |(f(x_u) - f(x_v)) - y_{u,v}|^\beta, & \text{otherwise.} \end{cases} \tag{3.22}$$

The SVR loss is defined as follows:

$$L_{SVR}(f; x_u, x_v, y_{u,v}) = \begin{cases} 0, & |y_{u,v} - (f(x_u) - f(x_v))| < \varepsilon, \\ |(f(x_u) - f(x_v)) - y_{u,v} - \varepsilon|^\beta, & \text{otherwise.} \end{cases} \tag{3.23}$$

The differences between the three loss functions lie in the different conditions that they penalize a mis-ranked pair (but not to what degree). For example, for the MP loss, not only the mis-ranked pairs but also the correctly-ranked pairs will be penalized if their magnitude of the predicted preference is too large; for the HMP loss, only the miss-ranked pairs are penalized; for the SVR loss, only if the magnitude of the predicted preference is different from the labeled preferences to a certain degree, the pair (no matter correctly or mis-ranked) will be penalized.

Then a L_2 regularization term is introduced to these loss functions, and the loss functions are optimized using kernel methods. Experimental results show that the magnitude-preserving loss functions can lead to better ranking performances than the original pairwise ranking algorithms, such as RankBoost [18].

3.3.3 IR-SVM

According to the second problem of the pairwise approach as mentioned above, the difference in the numbers of document pairs of different queries is usually significantly larger than the difference in the number of documents. This phenomenon has been observed in some previous studies [9, 27].

In this case, the pairwise loss function will be dominated by the queries with a large number of document pairs, and as a result the pairwise loss function will become inconsistent with the query-level evaluation measures. To tackle the problem, Cao et al. [9] propose introducing query-level normalization to the pairwise loss function. That is, the pairwise loss for a query will be normalized by the total number of document pairs associated with that query. In this way, the normalized pairwise losses with regards to different queries will become comparable to each other in their magnitude, no matter how many document pairs they are originally associated with. With this kind of query-level normalization, Ranking SVM will become a new algorithm, referred to as IR-SVM [9]. Specifically, given n training queries $\{q_i\}_{i=1}^n$, their associated document pairs $(x_u^{(i)}, x_v^{(i)})$, and the corresponding

Fig. 3.9 The sigmoid
function based loss ($\sigma = 5$)

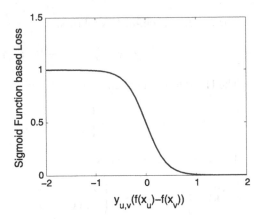

relevance judgment $y_{u,v}^{(i)}$, IR-SVM solves the following optimization problem,

$$\min \frac{1}{2}\|w\|^2 + \lambda \sum_{i=1}^{n} \frac{\sum_{u,v:y_{u,v}^{(i)}=1} \xi_{u,v}^{(i)}}{\tilde{m}^{(i)}}$$

$$\text{s.t.} \quad w^T\left(x_u^{(i)} - x_v^{(i)}\right) \geq 1 - \xi_{u,v}^{(i)}, \quad \text{if } y_{u,v}^{(i)} = 1 \qquad (3.24)$$

$$\xi_{u,v}^{(i)} \geq 0; \quad i = 1, \ldots, n;$$

where $\tilde{m}^{(i)}$ is the number of document pairs associated with query q_i.

According to the experimental results in [9], a significant performance improvement has been observed after the query-level normalization is introduced.

3.3.4 Robust Pairwise Ranking with Sigmoid Functions

In [10], Carvalho et al. try to tackle the third problem with the pairwise approach as aforementioned, i.e., the sensitivity to noisy relevance judgments.

Using some case studies, Carvalho et al. point out that the problem partly comes from the shapes of the loss functions. For example, in Ranking SVM [22, 23], the hinge loss is used. Due to its shape, outliers that produce large negative scores will have a strong contribution to the overall loss. In turn, these outliers will play an important role in determining the final learned ranking function.

To solve the problem, a new loss based on the sigmoid function (denoted as the sigmoid loss) is proposed to replace the hinge loss for learning to rank. Specifically the sigmoid loss has the following form (see Fig. 3.9):

$$L(f; x_u, x_v, y_{u,v}) = \frac{1}{1 + e^{-\sigma y_{u,v}(f(x_u) - f(x_v))}}. \qquad (3.25)$$

While the new loss function solves the problem of the hinge function, it also introduces a new problem. That is, the sigmoid loss is non-convex and thus the op-

Fig. 3.10 The ramp loss function

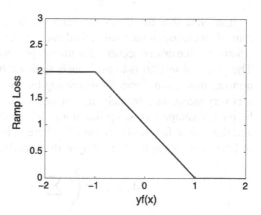

timization of it will be easily trapped into a local optimum. To tackle this challenge, the new ranker is used as a second optimization step, refining the ranking model produced by another ranker (e.g., Ranking SVM [22, 23]). In this way, with a relatively good starting point, it is more likely that the optimization of the sigmoid loss will lead to a reliably good solution.

The experimental results in [10] show that by using the new loss function, a better ranking performance can be achieved.

Actually in recent years, researchers have used similar ideas to improve the accuracy of classification. For example, the ramp loss (as shown in Fig. 3.10) is used in [13]. Like the sigmoid loss, the ramp loss also restricts the exceptional loss on outliers. According to [13], by using the ramp loss in SVMs, the number of support vectors can be significantly reduced. Further, considering the results presented in [3], better generalization ability can be achieved in this way.

3.3.5 P-norm Push

Although by looking at only a pair of documents one cannot determine their rank positions in the final ranked list, one can make a reasonable estimation on the rank position of a document by checking all the pairs containing the document. Since top positions are important for users, one can punish those errors occurring at top positions based on the estimation. In this way, one may be able to solve the forth problem with the pairwise approach.

Actually this is exactly the key idea in [29]. We refer to this algorithm as P-norm push. Suppose the pairwise loss function is $L(f; x_u, x_v, y_{u,v})$, then for document x_v, the overall error that it is mis-ranked before other documents can be written as follows:

$$L(f; x_v) = \sum_{u, y_{u,v}=1} L(f; x_u, x_v, y_{u,v}). \tag{3.26}$$

It is clear that the larger the value is, the more likely x_v is irrelevant but still ranked on the top of the final ranked list. As mentioned before, we should punish this situation since errors occurring at the top positions will damage the user experience. The proposal in [29] is to push such x_v down from the top. Specifically, a convex, non-negative, monotonically increasing function g is introduced for this purpose. If g is very steep, an extremely high price for a high-ranked irrelevant documents will be paid. Examples of steep functions include the exponential function ($g(r) = e^r$) and the power function ($g(r) = r^p$, where p is large). The power function is used in [29] as an example. Accordingly, the overall loss with regards to x_v becomes

$$L(f; x_v) = \left(\sum_{u, y_{u,v}=1} L(f; x_u, x_v, y_{u,v}) \right)^p. \tag{3.27}$$

After formulating the above loss function, a Boosting-style algorithm is developed to minimize the loss. Experimental results show that by introducing the extra penalty for the top-ranked irrelevant documents the ranking performance can be significantly improved.

3.3.6 Ordered Weighted Average for Ranking

In [34], Usunier et al. also aim at solving the forth problem with the pairwise approach. In particular, a weighted pairwise classification method is proposed, which emphasizes the correct ranking at the top positions of the final ranked list. Specifically, convex Ordered Weighted Averaging (OWA) operators [35] are used, which are parameterized by a set of decreasing weights α_t (the weight α_t is associated to the tth highest loss), to make a weighted sum of the pairwise classification losses. By choosing appropriate weights the convex OWA operators can lead to a loss function that focuses on the top-ranked documents.

Specifically, for $\forall s = (s_1, \ldots, s_m)$, its corresponding OWA is defined as follows:

$$owa^\alpha(s) = \sum_{t=1}^{m} \alpha_t s_{\sigma(t)}, \tag{3.28}$$

where σ is a permutation such that $\forall t, s_{\sigma(t)} \geq s_{\sigma(t+1)}$.

And accordingly, the loss functions in the pairwise ranking algorithms can be refined as follows.

$$L(f; x_u) = owa^\alpha_{v, y_{u,v}=1} \big(L(f; y_{u,v}, x_u, x_v) \big). \tag{3.29}$$

In general, the above loss associates the largest weights to the largest pairwise losses, and thus to the document pairs containing the irrelevant documents with the greatest relevance scores.

Experimental results demonstrate that the use of the OWA based loss can lead to better performance than the use of the original pairwise loss functions.

3.3.7 LambdaRank

In [7], another way of introducing position-based weights to the pairwise loss function is proposed. In particular, the evaluation measures (which are position based) are directly used to define the gradient with respect to each document pair in the training process. The hope is that the corresponding implicit loss function can be more consistent with the way that we evaluate the final ranking results.

Here we use an example to illustrate the basic idea. Suppose we have only two relevant documents x_1 and x_2, and their current ranks are as shown in Fig. 3.11. Suppose we are using NDCG@1 as the evaluation measure. Then, it is clear that if we can move either x_1 or x_2 to the top position of the ranked list, we will achieve the maximum NDCG@1.

It is clearly more convenient to move x_1 up since the effort will be much smaller than that for x_2. So, we can define (but not compute) the "gradient" with regards to the ranking score of x_1 (denoted as $s_1 = f(x_1)$) as larger than that with regards to the ranking score of x_2 (denoted as $s_2 = f(x_2)$). In other words, we can consider that there is an underlying implicit loss function L in the optimization process, which suggests

$$\frac{\partial L}{\partial s_1} > \frac{\partial L}{\partial s_2}. \tag{3.30}$$

The above "gradient" is called the lambda function, and this is why the algorithm is named LambdaRank. When we use NDCG to guide the training process, a specific form of the lambda function is given in [7]:

$$\lambda = Z_m \frac{2^{y_u} - 2^{y_v}}{1 + \exp(f(x_u) - f(x_v))} \big(\eta\big(r(x_u)\big) - \eta(x_v)\big), \tag{3.31}$$

where $r(\cdot)$ represents the rank position of a document in the previous iteration of training.

For each pair of documents x_u and x_v, in each round of optimization, their scores are updated by $+\lambda$ and $-\lambda$ respectively. In [15], the lambda functions for MAP and MRR are derived, and the local optimality of the LambdaRank algorithm is discussed. According to the experimental results, the LambdaRank algorithm can lead to local optimum of the objective function, and can have a competitive ranking performance.

Fig. 3.11 Optimizing
NDCG@1 by LambdaRank

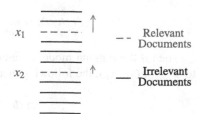

Note that in some previous work, LambdaRank is regarded as a kind of direct optimization method (see Sect. 4.2). However, according to our understanding, it is not such a method because the integral of the lambda function as defined above (which will correspond to the implicit loss function) is not highly related to NDCG. Therefore, in our opinion, it is more appropriate to say that LambdaRank considers the evaluation measures to set its pair weight than to say it directly optimizes the evaluation measures.

3.3.8 Robust Sparse Ranker

In [31], ranking is reduced to a weighted pairwise classification, whose weights are defined with the gain and discount functions in NDCG. In this way, one can also introduce positional information to the pairwise approach. Specifically, in [31], the following weighted classification loss is minimized:

$$L(c; \mathbf{x}, \mathbf{y}) = \sum_{u<v} \omega(y_u, y_v, c) \cdot \left[I_{\{y_u > y_v\}} \cdot c(x_u, x_v) \right], \tag{3.32}$$

where $\omega(y_u, y_v, c) = \frac{1}{Z_m} |(G(y_u) - G(y_v))(\eta(\pi(u)) - \eta(\pi(v)))|$ denotes the importance weight for the ordered pair (x_u, x_v), and c is a classifier: $c(x_u, x_v) = 1$ if c prefers x_v to x_u and 0 otherwise.

On the one hand, given any distribution P on the training data $\{\mathbf{x}, \mathbf{y}\}$, the expected loss of the classifier c is given by

$$L(c; P) = E_{\mathbf{x},\mathbf{y}} \left[L(c; \mathbf{x}, \mathbf{y}) \right]. \tag{3.33}$$

Then the regret of c on P is defined as follows:

$$r(c; P) = L(c; P) - \min_{c^*} L(c^*; P). \tag{3.34}$$

On the other hand, given the evaluation measure NDCG, we can also define its expectation with respect to the ranking model h as follows:

$$G(h; P) = E_{\mathbf{x},\mathbf{y}} \left[\text{NDCG}(h; \mathbf{x}, \mathbf{y}) \right]. \tag{3.35}$$

Correspondingly, we can define the regret of h on P as

$$r(h; P) = \max_{h^*} G(h^*; P) - G(h; P). \tag{3.36}$$

Then if the ranking model h determines the position of document x_u according to $\sum_{u \neq v} c(x_u, x_v)$, the following theoretical result has been proven in [31]:

$$r(h; P) < r(c; P'), \tag{3.37}$$

where P' denotes the induced distribution, defined by drawing an importance weighted pair from S which is randomly drawn from an underlying distribution P.

In other words, if we can minimize the regret of the classifier in the training process, correspondingly, we will also minimize the regret of the ranking model, and get NDCG effectively maximized.

Note that $\omega(y_u, y_v, c)$, the weight in the loss function (3.32), is a function of the classifier c. However, we actually learn this classifier by minimizing the loss function. Therefore, there is a chicken and egg problem in this process. To solve the problem, an online learning method is proposed. That is, one first sets an initial value c_0 for the classifier, and derives the way of updating the classifier (we denote the updated classifier as c_1) by minimizing the loss. After that, the loss function is updated by substituting c_1 into the weight. Then, the classifier is further updated to be c_2 by minimizing the updated loss function. This process is iterated until the process converges.

When applying the above idea to solve real ranking problems, a L_1 regularization term is added to the loss function, in order to guarantee the sparse solution of the parameter of the ranking model. According to the experimental results reported in [31], this robust sparse ranker can outperform many other ranking methods, including LambdaRank [7].

3.4 Summary

In this chapter, we have introduced several pairwise ranking methods, and shown how to improve their performances by balancing the distribution of document pairs across queries, emphasizing likely top-ranked documents, etc. Empirical studies have shown that such modifications really lead to performance gains.

In the next chapter, we will introduce the listwise approach, which also tries to solve the four problems with the pairwise approach, probably in a more natural and fundamental manner.

3.5 Exercises

3.1 What are the advantages of the pairwise approach as compared to the pointwise approach?

3.2 Study the relationship among the loss functions used in different pairwise ranking algorithms, and their relationships to the pairwise 0–1 loss.

3.3 Study the mathematical properties of the loss functions used in different pairwise ranking algorithms, such as the convexity.

3.4 Study the inter-dependency among document pairs, and discuss the potential issue of using the pairwise classification methods to perform learning to rank on such data.

3.5 Provide an intuitive explanation of the margin in Ranking SVM.

3.6 Ranking SVM can be extended to handle non-linear problems by using the kernel trick. Please give the corresponding deductions.

3.7 Derive the gradient of the loss function of RankNet with respect to the ranking model w (here we assume a linear ranking model is used).

3.8 Both P-Norm push and OWA target at emphasizing the top positions in the process of learning to rank. Please compare their behaviors by using an example, in a quantitative manner.

3.9 Since most pairwise ranking algorithms minimize the pairwise classification losses during training, while the ranking function (but not the pairwise classifier) is used in the test phase, there may be a theoretical gap between training and testing. Please take Ranking SVM as an example to show whether the gap exists and what the relationship between classification loss and ranking loss is.

References

1. Ahmad, N., Beg, M.: Fuzzy logic based rank aggregation methods for the World Wide Web. In: Proceedings of the International Conference on Artificial Intelligence in Engineering and Technology (ICAIET 2002), pp. 363–368 (2002)
2. Aslam, J.A., Montague, M.: Models for metasearch. In: Proceedings of the 24th Annual International ACM SIGIR Conference on Research and Development in Information Retrieval (SIGIR 2001), pp. 276–284 (2001)
3. Bax, E.: Nearly uniform validation improves compression-based error bounds. Journal of Machine Learning Research 9, 1741–1755 (2008)
4. Beg, M.M.S.: Parallel rank aggregation for the World Wide Web. World Wide Web Journal 6(1), 5–22 (2004)
5. Borda, J.: Mémoire sur les élections au scrutin. In: Histoire de l'Académie royale des sciences, pp. 42–51 (1781)
6. Bradley, R., Terry, M.: The rank analysis of incomplete block designs: the method of paired comparisons. Biometrika 39, 324–345 (1952)
7. Burges, C.J., Ragno, R., Le, Q.V.: Learning to rank with nonsmooth cost functions. In: Advances in Neural Information Processing Systems 19 (NIPS 2006), pp. 395–402 (2007)
8. Burges, C.J., Shaked, T., Renshaw, E., Lazier, A., Deeds, M., Hamilton, N., Hullender, G.: Learning to rank using gradient descent. In: Proceedings of the 22nd International Conference on Machine Learning (ICML 2005), pp. 89–96 (2005)
9. Cao, Y., Xu, J., Liu, T.-Y., Li, H., Huang, Y., Hon, H.-W.: Adapting ranking SVM to document retrieval. In: Proceedings of the 29th Annual International ACM SIGIR Conference on Research and Development in Information Retrieval (SIGIR 2006), pp. 186–193 (2006)
10. Carvalho, V.R., Elsas, J.L., Cohen, W.W., Carbonell, J.G.: A meta-learning approach for robust rank learning. In: SIGIR 2008 Workshop on Learning to Rank for Information Retrieval (LR4IR 2008) (2008)
11. Chapelle, O., Keerthi, S.S.: Efficient algorithms for ranking with SVMs. Information Retrieval Journal. Special Issue on Learning to Rank 13(3), doi:10.1007/s10791-009-9109-9 (2010)
12. Cohen, W.W., Schapire, R.E., Singer, Y.: Learning to order things. In: Advances in Neural Information Processing Systems (NIPS 1997), vol. 10, pp. 243–270 (1998)
13. Collobert, R., Sinz, F., Weston, J., Bottou, L.: Trading convexity for scalability. In: Proceedings of the 23rd International Conference on Machine Learning (ICML 2006), pp. 201–208. ACM, New York (2006). http://doi.acm.org/10.1145/1143844.1143870
14. Cortes, C., Mohri, M., et al.: Magnitude-preserving ranking algorithms. In: Proceedings of the 24th International Conference on Machine Learning (ICML 2007), pp. 169–176 (2007)

15. Donmez, P., Svore, K.M., Burges, C.J.C.: On the local optimality of lambdarank. In: Proceedings of the 32nd Annual International ACM SIGIR Conference on Research and Development in Information Retrieval (SIGIR 2009), pp. 460–467 (2009)
16. Dwork, C., Kumar, R., Naor, M., Sivakumar, D.: Rank aggregation methods for the web. In: Proceedings of the 10th International Conference on World Wide Web (WWW 2001), pp. 613–622. ACM, New York (2001)
17. Fagin, R., Kumar, R., Sivakumar, D.: Efficient similarity search and classification via rank aggregation. In: Proceedings of the 2003 ACM SIGMOD International Conference on Management of Data (SIGMOD 2003), pp. 301–312. ACM, New York (2003). http://doi.acm.org/10.1145/872757.872795
18. Freund, Y., Iyer, R., Schapire, R., Singer, Y.: An efficient boosting algorithm for combining preferences. Journal of Machine Learning Research 4, 933–969 (2003)
19. Freund, Y., Schapire, R.E.: A decision-theoretic generalization of online learning and an application to boosting. Journal of Computer and System Sciences 55(1), 119–139 (1995)
20. Friedman, J.: Greedy function approximation: a gradient boosting machine. Annual Statistics 29, 1189–1232 (2001)
21. Gao, J., Qi, H., Xia, X., Nie, J.: Linear discriminant model for information retrieval. In: Proceedings of the 28th Annual International ACM SIGIR Conference on Research and Development in Information Retrieval (SIGIR 2005), pp. 290–297 (2005)
22. Herbrich, R., Obermayer, K., Graepel, T.: Large margin rank boundaries for ordinal regression. In: Advances in Large Margin Classifiers, pp. 115–132 (2000)
23. Joachims, T.: Optimizing search engines using clickthrough data. In: Proceedings of the 8th ACM SIGKDD International Conference on Knowledge Discovery and Data Mining (KDD 2002), pp. 133–142 (2002)
24. Joachims, T.: Training linear svms in linear time. In: Proceedings of the 12th ACM SIGKDD International Conference on Knowledge Discovery and Data Mining (KDD 2006), pp. 217–226 (2006)
25. Matveeva, I., Burges, C., Burkard, T., Laucius, A., Wong, L.: High accuracy retrieval with multiple nested ranker. In: Proceedings of the 29th Annual International ACM SIGIR Conference on Research and Development in Information Retrieval (SIGIR 2006), pp. 437–444 (2006)
26. Qin, T., Liu, T.-Y., Lai, W., Zhang, X.-D., Wang, D.-S., Li, H.: Ranking with multiple hyperplanes. In: Proceedings of the 30th Annual International ACM SIGIR Conference on Research and Development in Information Retrieval (SIGIR 2007), pp. 279–286 (2007)
27. Qin, T., Zhang, X.-D., Tsai, M.-F., Wang, D.-S., Liu, T.-Y., Li, H.: Query-level loss functions for information retrieval. Information Processing and Management 44(2), 838–855 (2008)
28. Rigutini, L., Papini, T., Maggini, M., Scarselli, F.: Learning to rank by a neural-based sorting algorithm. In: SIGIR 2008 Workshop on Learning to Rank for Information Retrieval (LR4IR 2008) (2008)
29. Rudin, C.: Ranking with a p-norm push. In: Proceedings of the 19th Annual Conference on Learning Theory (COLT 2006), pp. 589–604 (2006)
30. Shen, L., Joshi, A.K.: Ranking and reranking with perceptron. Journal of Machine Learning 60(1–3), 73–96 (2005)
31. Sun, Z., Qin, T., Tao, Q., Wang, J.: Robust sparse rank learning for non-smooth ranking measures. In: Proceedings of the 32nd Annual International ACM SIGIR Conference on Research and Development in Information Retrieval (SIGIR 2009), pp. 259–266 (2009)
32. Thurstone, L.: A law of comparative judgement. Psychological Review 34, 34 (1927)
33. Tsai, M.-F., Liu, T.-Y., Qin, T., Chen, H.-H., Ma, W.-Y.: Frank: a ranking method with fidelity loss. In: Proceedings of the 30th Annual International ACM SIGIR Conference on Research and Development in Information Retrieval (SIGIR 2007), pp. 383–390 (2007)
34. Usunier, N., Buffoni, D., Gallinari, P.: Ranking with ordered weighted pairwise classification. In: Proceedings of the 26th International Conference on Machine Learning (ICML 2009), pp. 1057–1064 (2009)
35. Yager, R.R.: On ordered weighted averaging aggregation operators in multi-criteria decision making. IEEE Transactions on Systems, Man, and Cybernetics 18(1), 183–190 (1988)

36. Zheng, Z., Chen, K., Sun, G., Zha, H.: A regression framework for learning ranking functions using relative relevance judgments. In: Proceedings of the 30th Annual International ACM SIGIR Conference on Research and Development in Information Retrieval (SIGIR 2007), pp. 287–294 (2007)
37. Zheng, Z., Zha, H., Sun, G.: Query-level learning to rank using isotonic regression. In: SIGIR 2008 Workshop on Learning to Rank for Information Retrieval (LR4IR 2008) (2008)
38. Zheng, Z., Zha, H., Zhang, T., Chapelle, O., Chen, K., Sun, G.: A general boosting method and its application to learning ranking functions for web search. In: Advances in Neural Information Processing Systems 20 (NIPS 2007), pp. 1697–1704 (2008)
39. Zhou, K., Xue, G.-R., Zha, H., Yu, Y.: Learning to rank with ties. In: Proceedings of the 31st Annual International ACM SIGIR Conference on Research and Development in Information Retrieval (SIGIR 2008), pp. 275–282 (2008)

Chapter 4
The Listwise Approach

Abstract In this chapter, we introduce the listwise approach to learning to rank. Specifically, we first introduce those listwise ranking algorithms that minimize measure-specific loss functions (also referred to as direct optimization methods), and then introduce some other algorithms whose loss functions are not directly related to evaluation measures for information retrieval.

4.1 Overview

The listwise approach takes the entire set of documents associated with a query in the training data as the input and predicts their ground truth labels. According to the loss functions used, the approach can be divided into two sub-categories. For the first sub-category, the loss function is explicitly related to evaluation measures (e.g., approximation or upper bound of the measure-based ranking errors). Example algorithms include [5, 7, 25, 27, 32–34, 36, 37]. Due to the strong relationship between the loss functions and the evaluation measures, these algorithms are also referred to as direct optimization methods. For the second sub-category, the loss function is not explicitly related to the evaluation measures. Example algorithms include [4, 16, 30, 31]. In the rest of this chapter, we will introduce both sub-categories and their representative algorithms.

Note that the listwise approach assumes that the ground truth labels are given in terms of permutations, while the judgments might be in other forms (e.g., relevance degrees or pairwise preferences). In this case, we will use the concept of equivalent permutation set (denoted as Ω_y, see Chap. 1) to bridge the gap. With Ω_y, we use $L(f; \mathbf{x}, \Omega_y)$ to represent the loss function for the listwise approach. However, for ease of discussion, we sometimes still write the loss function as $L(f; \mathbf{x}, \mathbf{y})$ if Ω_y is generated from the relevance degree of each single document.

Furthermore, evaluation measures such as MAP and NDCG can also be rewritten in the form of Ω_y. For example, we can rewrite NDCG as follows:

$$\text{NDCG}(\pi, \Omega_y) = \frac{1}{Z_m} \sum_{j=1}^{m} G\big(l_{\pi^{-1}(\pi_y(j))}\big) \eta\big(\pi_y(j)\big), \quad \forall \pi_y \in \Omega_y. \quad (4.1)$$

T.-Y. Liu, *Learning to Rank for Information Retrieval*,
DOI 10.1007/978-3-642-14267-3_4, © Springer-Verlag Berlin Heidelberg 2011

Again, for ease of discussion, we sometimes also write NDCG as $\text{NDCG}(\pi, \mathbf{y})$ or $\text{NDCG}(f, \mathbf{x}, \mathbf{y})$ (when $\pi = sort \circ f(\mathbf{x})$).

4.2 Minimization of Measure-Specific Loss

It might be the most straightforward choice to learn the ranking model by directly optimizing what is used to evaluate the ranking performance. This is exactly the motivation of the first sub-category of the listwise approach, i.e., the direct optimization methods. However, the task is not as trivial as it seems. As mentioned before, evaluation measures such as NDCG and MAP are position based, and thus discontinuous and non-differentiable [26, 33]. The difficulty of optimizing such objective functions stems from the fact that most existing optimization techniques were developed to handle continuous and differentiable cases.

To tackle the challenges, several attempts have been made. First, one can choose to optimize a continuous and differentiable approximation of the measure-based ranking error. By doing so, many existing optimization technologies can be leveraged. Example algorithms include SoftRank [27], Approximate Rank [25], and SmoothRank [7]. Second, one can alternatively optimize a continuous and differentiable (and sometimes even convex) bound of the measure-based ranking error. Example algorithms include SVM^{map} [36], SVM^{ndcg} [5], and PermuRank [33].[1] Third, one can choose to use optimization technologies that are able to optimize non-smooth objectives. For example, one can leverage the Boosting framework for this purpose (the corresponding algorithm is called AdaRank [32]), or adopt the genetic algorithm for the optimization (the corresponding algorithm is called RankGP [34]).

4.2.1 Measure Approximation

4.2.1.1 SoftRank

In SoftRank [27], it is assumed that the ranking of the documents is not simply determined by sorting according to the scoring function. Instead, it introduces randomness to the ranking process by regarding the real score of a document as a random variable whose mean is the score given by the scoring function. In this way, it is possible that a document can be ranked at any position, of course with different probabilities. For each such possible ranking, one can compute an NDCG value. Then the expectation of NDCG over all possible rankings can be used as an smooth approximation of the original evaluation measure NDCG. The detailed steps to achieve this goal are elaborated as follows.

---

[1] Actually, this trick has also been used in classification. That is, since the 0–1 classification loss is non-differentiable, convex upper bounds like the exponential loss have been used instead.

First, one defines the score distribution. Given $\mathbf{x} = \{x_j\}_{j=1}^m$ associated with a training query q, the score s_j of document x_j is treated as no longer a deterministic value but a random variable. The random variable is governed by a Gaussian distribution whose variance is σ_s and mean is $f(x_j)$, the original score outputted by the scoring function. That is,

$$p(s_j) = N\left(s_j | f(x_j), \sigma_s^2\right). \tag{4.2}$$

Second, one defines the rank distribution. Due to the randomness in the scores, every document has the probability of being ranked at any position. Specifically, based on the score distribution, the probability of a document x_u being ranked before another document x_v can be deduced as follows:

$$P_{u,v} = \int_0^\infty N\left(s | f(x_u) - f(x_v), 2\sigma_s^2\right) ds. \tag{4.3}$$

On this basis, the rank distribution can be derived in an iterative manner, by adding the documents into the ranked list one after another. Suppose we already have document x_j in the ranked list, when adding document x_u, if document x_u can beat x_j the rank of x_j will be increased by one. Otherwise the rank of x_j will remain unchanged. Mathematically, the probability of x_j being ranked at position r (denoted as $P_j(r)$) can be computed as follows:

$$P_j^{(u)}(r) = P_j^{(u-1)}(r-1)P_{u,j} + P_j^{(u-1)}(r)(1 - P_{u,j}). \tag{4.4}$$

Third, with the rank distribution, one computes the expectation of NDCG (referred to as SoftNDCG[2]), and use $(1 - \text{SoftNDCG})$ as the loss function in learning to rank:

$$L(f; \mathbf{x}, \mathbf{y}) = 1 - \frac{1}{Z_m} \sum_{j=1}^m (2^{y_j} - 1) \sum_{r=1}^m \eta(r) P_j(r). \tag{4.5}$$

In order to learn the ranking model f by minimizing the above loss function, one can use a neural network as the model, and taking gradient descent as the optimization algorithm.

In [14], the Gaussian process is used to further enhance the SoftRank algorithm, where σ_s is no longer a pre-specified constant but a learned parameter. In [35], the SoftRank method is further extended to approximate $P@k$ and AP. The corresponding objective functions are named SoftPC and SoftAP respectively.

4.2.1.2 Decision Theoretic Framework for Ranking

In [37], a similar idea to that of SoftRank is proposed, which uses a decision theoretic framework to optimize expected evaluation measures. First, a ranking utility is

---

[2]For ease of reference, we also refer to objective functions like SoftNDCG as surrogate measures.

defined. For example, the utility can be any evaluation measure. Suppose the utility
is denoted as $U(\cdot)$, then the following expected utility is used as the objective for
learning:

$$\tilde{U}(w; \mathbf{x}, \mathbf{y}) = E_{p(\mathbf{s}|\mathbf{x},w)}\big[U(\mathbf{s}, \mathbf{y})\big], \qquad (4.6)$$

where w is the parameter of the ranking function, and $\mathbf{s}$ is the ranking scores of the
documents.

With the above definition, although the utility itself may be discontinuous and
non-differentiable, it does not contain any model parameter and will not bring trou-
ble to the learning process. What contains the model parameter is the conditional
probability $p(\mathbf{s}|\mathbf{x}, w)$.

In [37], the independence between the documents (i.e., $p(\mathbf{s}|\mathbf{x}, w) =$
$\prod_j p(s_j|x_j, w)$) is assumed, and a generalized linear model [22] is used to defined
$p(s_j|x_j, w)$, the conditional probability for each single document. A generalized
linear model consists of a likelihood model $p(y|\theta)$, a linear combination of the in-
puts and model parameters $w^T x$, and a link function that maps the parameter θ to
the real line. In particular, a binomial function (i.e., Bin$(\cdot)$) is used as the likelihood
model in [37], and a cumulative normal function (i.e., $\Psi(\cdot)$) is used to define the
probit link function:

$$p(s_j|\mathbf{x}, w) = \text{Bin}\left(s_j; \Psi\left(w^T x_j; 0, \frac{1}{\pi}\right), m\right), \qquad (4.7)$$

where m is the number of documents.

Then the algorithm maximizes the expected utility (equivalently, the negative
utility can be regarded as the loss function). In particular, the approximate Bayesian
inference procedure [18] with a factorized Gaussian prior is used to learn the model
parameters.

Note that in addition to optimizing evaluation measures, the decision theoretical
framework can also be used to optimize other utilities, e.g., the probability of user
clicks on the documents.

4.2.1.3 Approximate Rank

In [25], Qin et al. point out that the underlying reason why evaluation measures
are non-smooth is that rank positions are non-smooth with respect to ranking
scores. Therefore, they propose performing approximation to the rank positions using smooth functions of the ranking scores, such that the approximate evaluation
measures can consequently become differentiable and easier to optimize.

Here we take NDCG as an example to illustrate the approximation process. The
same idea also applies to MAP. For more details, please refer to [25].

If one changes the index of summation in the definition of NDCG (see Chap. 1)
from the positions in the ranking result to the indexes of documents, NDCG can be

rewritten as[3]

$$Z_m^{-1} \sum_j \frac{G(y_j)}{\log(1 + \pi(x_j))}, \tag{4.8}$$

where $\pi(x_j)$ is the position of document x_j in the ranking result π, which can be calculated as

$$\pi(x_j) = 1 + \sum_{u \neq j} I_{\{f(x_j) - f(x_u) < 0\}}. \tag{4.9}$$

From the above equation, one can clearly see where the non-smooth nature of NDCG comes from. Actually, NDCG is a smooth function of the rank position, however, the rank position is a non-smooth function of the ranking scores because of the indicator function.

The key idea in [25] is to approximate the indicator function by a sigmoid function,[4] such that the position can be approximated by a smooth function of the ranking scores:

$$\widehat{\pi}(x_j) = 1 + \sum_{u \neq j} \frac{\exp(-\alpha(f(x_j) - f(x_u)))}{1 + \exp(-\alpha(f(x_j) - f(x_u)))}, \tag{4.10}$$

where $\alpha > 0$ is a scaling constant.

By substituting $\pi(x_j)$ in (4.8) by $\widehat{\pi}(x_j)$, one can get the approximation for NDCG (denoted as AppNDCG), and then define the loss function as $(1 - \text{AppNDCG})$,

$$L(f; \mathbf{x}, \mathbf{y}) = 1 - Z_m^{-1} \sum_{j=1}^{m} \frac{G(y_j)}{\log(1 + \widehat{\pi}(x_j))}. \tag{4.11}$$

Since this loss function is continuous and differentiable with respect to the scoring function, one can simply use the gradient descent method to minimize it.

The approximation accuracy is analyzed in [25]. The basic conclusion is that when α is set to be sufficiently large, the approximation can become very accurate. In other words, if one can find the optimum of the proposed loss function, it is very likely one also obtains the optimum of NDCG.

Note that the more accurate the approximation is, the more "non-smooth" the loss function becomes. Correspondingly, the optimization process becomes less robust. Usually one needs to adopt some robust optimization algorithm such as simulated annealing or random optimization to ensure the finding of a good solution to the optimization problem.

---

[3]Note that here we directly assume $\eta(r) = \frac{1}{\log(1+r)}$ for simplicity although other forms of discount function are also valid in defining NDCG.

[4]Note that sigmoid function is a large family of functions. For example the widely used logistic function is a special case of the sigmoid function, and other family members include the ordinary arc-tangent function, the hyperbolic tangent function and the error function. Here the logistic function is taken as an example of derivation.

4.2.1.4 SmoothRank

In [7], a similar idea to Approximate Rank is proposed. That is, the evaluation measures are smoothed by approximating the rank position. The slight difference lies in the way of defining the approximator and the way of performing the optimization.

Again, here we use NDCG as an example for illustration. Please note that the same idea also applies to MAP and many other evaluation measures.

In [7], NDCG is rewritten by using both the indexes of the documents and the positions in the ranking result:

$$\sum_{j=1}^{m} \sum_{u=1}^{m} G(y_{\pi^{-1}(u)}) \eta(u) I_{\{x_j = x_{\pi^{-1}(u)}\}}, \tag{4.12}$$

where $I_{\{x_j = x_{\pi^{-1}(u)}\}}$ indicates whether document x_j is ranked at the uth position.

Then the following soft version of the indicator function $I_{\{x_j = x_{\pi^{-1}(u)}\}}$ is introduced:

$$h_{ju} = \frac{e^{-(f(x_j) - f(x_{\pi^{-1}(u)}))^2/\sigma}}{\sum_{l=1}^{m} e^{-(f(x_l) - f(x_{\pi^{-1}(u)}))^2/\sigma}}. \tag{4.13}$$

With h_{ju}, one can get the smoothened version of NDCG, and define the loss function on its basis:

$$L(f; \mathbf{x}, \mathbf{y}) = 1 - \sum_{j=1}^{m} \sum_{u=1}^{m} G(y_{\pi^{-1}(u)}) \eta(u) h_{ju}$$

$$= 1 - \sum_{j=1}^{m} \sum_{u=1}^{m} G(y_{\pi^{-1}(u)}) \eta(u) \frac{e^{-(f(x_j) - f(x_{\pi^{-1}(u)}))^2/\sigma}}{\sum_{l=1}^{m} e^{-(f(x_l) - f(x_{\pi^{-1}(u)}))^2/\sigma}}. \tag{4.14}$$

It is clear that h_{ju} is a continuous function of the scoring function f, and so is the above loss function. Therefore, one can use the gradient descent method to optimize it.

Similar to the discussions about Approximate Rank, the choice of the smoothing parameter σ is critical: on the one hand, if it is small the objective function will be very close to the original evaluation measure and is therefore highly non-smooth and difficult to optimize; on the other hand, if it is large, the objective function is smooth and easy to optimize, but substantially different from the corresponding evaluation measure. In [7], a deterministic annealing strategy is used to assist the optimization. This procedure can be seen as an homotopy method where a series of functions are constructed: the first one is easy to optimize, the last one is the function of interest and each function in the middle can be seen as a deformed function of the previous one. The homotopy method iteratively minimizes each of these functions starting from the minimizer of the previous function. In particular, the deformation is controlled by σ: one starts with a large σ, and the minimizer iteratively reduces σ by steps.

4.2.2 Bound Optimization

4.2.2.1 SVM^{map}

SVM^{map} [36] uses the framework of structured SVM [17, 29] to optimize the evaluation measure AP.

Suppose $\mathbf{x} = \{x_j\}_{j=1}^m$ represents all the documents associated with the training query q, $\mathbf{y} = \{y_j\}_{j=1}^m$ ($y_j = 1$, if document x_j is labeled as relevant; $y_j = 0$, otherwise) represents the corresponding ground truth label, and $\mathbf{y}^c$ represents any incorrect label. Then SVM^{map} is formulated as follows:

$$\min \frac{1}{2}\|w\|^2 + \frac{\lambda}{n}\sum_{i=1}^{n}\xi^{(i)}$$

$$\text{s.t.} \quad \forall \mathbf{y}^{c(i)} \neq \mathbf{y}^{(i)}, \tag{4.15}$$

$$w^T\Psi\left(\mathbf{y}^{(i)},\mathbf{x}^{(i)}\right) \geq w^T\Psi\left(\mathbf{y}^{c(i)},\mathbf{x}^{(i)}\right) + 1 - \text{AP}\left(\mathbf{y}^{c(i)},\mathbf{y}^{(i)}\right) - \xi^{(i)}.$$

Here Ψ is called the joint feature map, whose definition is given as

$$\Psi(\mathbf{y},\mathbf{x}) = \sum_{u,v:y_u=1,y_v=0}(x_u - x_v), \tag{4.16}$$

$$\Psi\left(\mathbf{y}^c,\mathbf{x}\right) = \sum_{u,v:y_u=1,y_v=0}\left(y_u^c - y_v^c\right)(x_u - x_v). \tag{4.17}$$

It has been proven that the sum of slacks in SVM^{map} can bound $(1 - \text{AP})$ from above. Therefore, if we can effectively solve the above optimization problem, AP will be correspondingly maximized. However, there are an exponential number of incorrect labels for the documents, and thus the optimization problem has an exponential number of constraints for each query. Therefore, it is a big challenge to directly solve such an optimization problem. To tackle the challenge, the active set method is used in [36]. That is, a working set is maintained, which only contains those constraints with the largest violations (defined below), and the optimization is performed only with respect to this working set.

$$\text{Violation} \triangleq 1 - \text{AP}\left(\mathbf{y}^c,\mathbf{y}\right) + w^T\Psi\left(\mathbf{y}^c,\mathbf{x}\right). \tag{4.18}$$

Then the problem becomes efficiently finding the most violated constraints for a given scoring function $f(x) = w^Tx$. To this end, the property of AP is considered. That is, if the relevance at each position is fixed, AP will be the same no matter which document appears at that position. Accordingly, an efficient strategy to find the most violated constraint can be designed [36], whose time complexity is only $O(m\log m)$, where m is the number of documents associated with query q.

In [5, 6, 19], SVM^{map} is further extended to optimize other evaluation measures. Specifically, when the target evaluation measures are NDCG and MRR, the resultant algorithms are referred to as SVM^{ndcg} and SVM^{mrr}. Basically, in these extensions, different feature maps or different strategies of searching the most violated constraints are used, but the key idea remains the same as that of SVM^{map}.

Further analysis in [33] shows that the loss functions in the aforementioned works are actually convex upper bounds of the following quantity, and this quantity is an upper bound of the corresponding measure-based ranking error.

$$\max_{\mathbf{y}^c \neq \mathbf{y}}\big(1 - M\big(\mathbf{y}^c, \mathbf{y}\big)\big)I_{\{w^T \Psi(\mathbf{y},\mathbf{x}) \leq w^T \Psi(\mathbf{y}^c,\mathbf{x})\}}, \tag{4.19}$$

where $M(\mathbf{y}^c, \mathbf{y})$ represents the value of an evaluation measure M when the ranking result is $\mathbf{y}^c$ and the ground truth label is $\mathbf{y}$.

In [5, 6, 19, 36] the following convex upper bound of the above quantity is minimized:

$$\max_{\mathbf{y}^c \neq \mathbf{y}}\big[1 - M\big(\mathbf{y}^c, \mathbf{y}\big) + w^T \Psi\big(\mathbf{y}^c, \mathbf{x}\big) - w^T \Psi(\mathbf{y}, \mathbf{x})\big]_+. \tag{4.20}$$

In another method, called PermuRank [33], a different convex upper bound of the aforementioned quantity is employed in the optimization process, as shown below.

$$\max_{\mathbf{y}^c \neq \mathbf{y}}\big(1 - M\big(\mathbf{y}^c, \mathbf{y}\big)\big)\big[1 - w^T \Psi(\mathbf{y}, \mathbf{x}) + w^T \Psi\big(\mathbf{y}^c, \mathbf{x}\big)\big]_+. \tag{4.21}$$

4.2.3 Non-smooth Optimization

Different from the methods introduced in the previous two subsections, there are also some other methods that directly optimize evaluation measures using non-smooth optimization techniques. For example, in AdaRank [32], the boosting algorithm is used to optimize the exponential function of the evaluation measure. Note that since the exponential function is monotonic, the optimization of the objective function in AdaRank is equivalent to the optimization of the evaluation measure itself. For another example, in RankGP [34], the evaluation measure is used as the fitness function of a genetic algorithm.

It should be noted that although these non-smooth optimization techniques are very general and can deal with the optimization of any non-smooth functions, they also have their limitations. That is, although the evaluation measures are used directly as the objective function, it is not guaranteed that one can really find their optima.

4.2.3.1 AdaRank

Xu and Li [32] point out that evaluation measures can be plugged into the framework of Boosting and get effectively optimized. This process does not require the

---

Algorithm 1: Learning algorithms for AdaRank

Input: the set of documents associated with each query
Given: initial distribution $\mathcal{D}_1$ on input queries
For $t = 1, \ldots, T$
    Train weak ranker $f_t(\cdot)$ based on distribution $\mathcal{D}_t$.
    Choose $\alpha_t = \frac{1}{2} \log \frac{\sum_{i=1}^{n} \mathcal{D}_t(i)(1+M(f_t, \mathbf{x}^{(i)}, \mathbf{y}^{(i)}))}{\sum_{i=1}^{n} \mathcal{D}_t(i)(1-M(f_t, \mathbf{x}^{(i)}, \mathbf{y}^{(i)}))}$
    Update $\mathcal{D}_{t+1}(i) = \frac{\exp(-M(\sum_{s=1}^{t} \alpha_s f_s, \mathbf{x}^{(i)}, \mathbf{y}^{(i)}))}{\sum_{j=1}^{n} \exp(-M(\sum_{s=1}^{t} \alpha_s f_s, \mathbf{x}^{(j)}, \mathbf{y}^{(j)}))}$.
Output: $\sum_t \alpha_t f_t(\cdot)$.

---

evaluation measures to be continuous and differentiable. The resultant algorithm is called AdaRank.

As we know, in the conventional AdaBoost algorithm, the exponential loss is used to update the distribution of input objects and to determine the combination coefficient of the weak learners at each round of iteration (see Chaps. 21 and 22). In AdaRank, evaluation measures are used to update the distribution of queries and to compute the combination coefficient of the weak rankers. The algorithm flow is shown in Algorithm 1, where $M(f, \mathbf{x}, \mathbf{y})$ represents the evaluation measure. Due to the analogy to AdaBoost, AdaRank can focus on those hard queries and progressively minimize $1 - M(f, \mathbf{x}, \mathbf{y})$.

The condition for the convergence of the training process is given in [32], with a similar derivation technique to that for AdaBoost. The condition requires $|M(\sum_{s=1}^{t} \alpha_s f_s, \mathbf{x}, \mathbf{y}) - M(\sum_{s=1}^{t-1} \alpha_s f_s, \mathbf{x}, \mathbf{y}) - \alpha_t M(f_t, \mathbf{x}, \mathbf{y})|$ to be very small. This implies the assumption on the linearity of the evaluation measure M, as a function of f_t. However, this assumption may not be well satisfied in practice. Therefore, it is possible that the training process of AdaRank does not naturally converge and some additional stopping criteria are needed.

4.2.3.2   Genetic Programming Based Algorithms

Genetic programming is a method designed for optimizing complex objectives. In the literature of learning to rank, there have been several attempts on using genetic programming to optimize evaluation measures. Representative algorithms include [1, 8–13, 28, 34].

Here we take the algorithm named RankGP [34] as an example to illustrate how genetic programming can be used to learn the ranking model. In this algorithm, the ranking model is defined as a tree, whose leaf nodes are features or constants, and non-leaf nodes are operators such as $+, -, \times, \div$ (see Fig. 4.1). Then single population genetic programming is used to perform the learning on the tree. Cross-over, mutation, reproduction, and tournament selection are used as evolution mechanisms, and the evaluation measure is used as the fitness function.

Fig. 4.1 Ranking function
used in RankGP

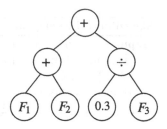

4.2.4 Discussions

As can be seen from the above examples, various methods have been proposed to
optimize loss functions that are related to the evaluation measures. Actually there
are even more such works, which have not been introduced in detail due to space
restrictions, such as [23, 26].

In addition to these works, in [35] it is discussed whether an evaluation measure
is suitable for direct optimization. For this purpose, a concept called informativeness is proposed. Empirical studies show that more informative measures can lead
to more effective ranking models. Discussions are then conducted on why some
measures are more informative while the others are less so. Specifically, according
to the discussions, multi-level measures (such as NDCG) are more informative than
binary measures (such as AP), since they respond to any flips between documents
of different relevance degrees. Furthermore, given two binary (or multi-level) measures, one measure may still be more informative than the other for the following
reasons: (1) some measures respond only to flips in some specific part of the ranking, usually the top end; (2) even if two measures depend on the same part of the
ranking, one may be insensitive to some of these flips within this part due to ignoring
some flips or due to the discount functions used.

The experiments with several direct optimization methods have verified the
above discussions. As an interesting result, it is not always the best choice to optimize a less informative measure even it is eventually used for evaluation. For example, directly optimizing AP can lead to a better ranking performance in terms of
P@10 than directly optimizing P@10.

4.3 Minimization of Non-measure-Specific Loss

In the second sub-category of the listwise approach, the loss function, which is not
measure specific, reflects the inconsistency between the output of the ranking model
and the ground truth permutation π_y. Although evaluation measures are not directly
optimized here, if one can consider the distinct properties of ranking in information
retrieval in the design of the loss function, it is also possible that the model learned
can have good performance in terms of evaluation measures.

Example algorithms in this sub-category include ListNet [4], ListMLE [31],
StructRank [16], and BoltzRank [30]. We will give introductions to them in this
section.

4.3.1 ListNet

In ListNet [4], the loss function is defined using the probability distribution on permutations.

Actually the probability distributions on permutations have been well studied in the field of probability theory. Many famous models have been proposed to represent permutation probability distributions, such as the Plackett–Luce model [20, 24] and the Mallows model [21]. Since a permutation has a natural one-to-one correspondence with a ranked list, these researches can be naturally applied to ranking. ListNet [4] is just such an example, demonstrating how to apply the Plackett–Luce model to learning to rank.

Given the ranking scores of the documents outputted by the scoring function f (i.e., $\mathbf{s} = \{s_j\}_{j=1}^m$, where $s_j = f(x_j)$), the Plackett–Luce model defines a probability for each possible permutation π of the documents, based on the chain rule, as follows:

$$P(\pi|\mathbf{s}) = \prod_{j=1}^m \frac{\varphi(s_{\pi^{-1}(j)})}{\sum_{u=1}^m \varphi(s_{\pi^{-1}(u)})}, \tag{4.22}$$

where $\pi^{-1}(j)$ denotes the document ranked at the jth position of permutation π and φ is a transformation function, which can be linear, exponential, or sigmoid.

Please note that the Plackett–Luce model is scale invariant and translation invariant in certain conditions. For example, when we use the exponential function as the transformation function, after adding the same constant to all the ranking scores, the permutation probability distribution defined by the Plackett–Luce model will not change. When we use the linear function as the transformation function, after multiplying all the ranking scores by the same constant, the permutation probability distribution will not change. These properties are quite in accordance with our intuitions on ranking.

With the Plackett–Luce model, for a given query q, ListNet first defines the permutation probability distribution based on the scores given by the scoring function f. Then it defines another permutation probability distribution $P_y(\pi)$ based on the ground truth label. For example, if the ground truth is given as relevance degrees, it can be directly substituted into the Plackett–Luce model to obtain a probability distribution. When the ground truth is given as a permutation π_y (or a permutation set Ω_y), one can define $P_y(\pi)$ as a delta function which only takes a value of 1 for this permutation (or the permutations in the set), and takes a value of 0 for all the other permutations. One can also first employ a mapping function to map the positions in the ground truth permutation to real-valued scores and then use (4.22) to compute the probability distribution. Furthermore, one can also choose to use the Mallows model [21] to define $P_y(\pi)$ by taking the ground truth permutation as the centroid in the model.

For the next step, ListNet uses the K-L divergence between the probability distribution for the ranking model and that for the ground truth to define its loss function (denoted as the *K-L divergence loss* for short).

$$L(f; \mathbf{x}, \Omega_y) = D\big(P_y(\pi) \,\|\, P\big(\pi \,|\, (f(w, \mathbf{x}))\big)\big). \tag{4.23}$$

It has been proven in [31] that the above K-L divergence loss function is convex, and therefore one can use a simple gradient descent method to get its global optimum. As shown in [4], the training curve of ListNet demonstrates well the correlation between the K-L divergence loss and $1 - \text{NDCG@5}$, although the loss is not explicitly related to NDCG.

As one may have noticed, there is a computational issue with the ListNet algorithm. Although due to the use of the scoring function to define the hypothesis, the testing complexity of ListNet can be the same as those of the pointwise and pairwise approaches, the training complexity of ListNet is much higher. The training complexity is in the exponential order of m (and thus intractable in practice), because the evaluation of the K-L divergence loss for each query q requires the addition of m-factorial terms. Comparatively speaking, the training complexities of the pointwise and pairwise approaches are roughly proportional to the number of documents (i.e., $O(m)$) and the number of document pairs (i.e., $O(\tilde{m})$). To tackle the problem, a top-k version of the K-L divergence loss is further introduced in [4], which is based on the top-k Plackett–Luce model and can reduce the training complexity from the exponential to the polynomial order of m.

4.3.2 ListMLE

Even if the top-k K-L divergence loss is used, one still cannot avoid its following limitations of ListNet. When k is set to be large, the time complexity of evaluating the K-L divergence loss is still very high. However, when k is set to be small, information about the permutation will be significantly lost and the effectiveness of the ListNet algorithm becomes not guaranteed [4].

To tackle the problem, a new algorithm, named ListMLE, is proposed [31]. ListMLE is also based on the Plackett–Luce model. For each query q, with the permutation probability distribution defined with the output of the scoring function, it uses the negative log likelihood of the ground truth permutation as the loss function.[5] We denote this new loss function as the *likelihood loss* for short:

$$L(f; \mathbf{x}, \pi_y) = -\log P\big(\pi_y | f(w, \mathbf{x})\big). \tag{4.24}$$

It is clear that in this way the training complexity has been greatly reduced as compared to ListNet, since one only needs to compute the probability of a single permutation π_y but not all the permutations. Once again, it can be proven that this

---

[5]Please note that one can also use the top-k Plackett–Luce model to define the likelihood loss. However, in this case, the purpose is not to reduce the computational complexity but to better reflect the real ranking requirements.

loss function is convex, and therefore one can safely use a gradient descent method to optimize the loss.

Here we would like to point out that ListMLE can be regarded as a special case of ListNet. If we set $P_y(\pi)$ in ListNet to be 1 for π_y, and 0 for all the other permutations, the K-L divergence between $P(\pi \mid f(w, \mathbf{x}))$ and $P_y(\pi)$ will turn out to be the negative log likelihood as defined in (4.24).

When the judgment is not total order, one needs to use the concept of equivalent permutation set Ω_y to extend the ListMLE algorithm. That is, we can regard all the permutations in Ω_y to be the ground truth permutations. Then we can choose to maximize the average likelihood of the permutations in Ω_y, or to maximize the largest likelihood in it (inspired by multi-instance learning [2]). In the latter case, the loss function becomes

$$L(f; \mathbf{x}, \Omega_y) = \min_{\pi_y \in \Omega_y} \left(-\log P\big(\pi_y \mid f(w, \mathbf{x})\big)\right). \qquad (4.25)$$

Since the size of the constraint set is exponentially large, it is costly to find the permutation with the smallest loss in each round of iteration. In [31] an effective way of conducting the search is proposed. That is, one can first sort the objects according to their ratings in the ground truth. Then for the objects having the same rating, one sorts them in descending order of their ranking scores. It has been proven that the resultant permutation has the smallest loss. Also, the minimal loss changes with respect to w, which suggests that the optimization process will be an iterative one. It has been proven that this iterative process can converge in polynomial time.

As an extension of ListMLE, in [15], a Gamma distribution based prior is introduced to regularize the likelihood loss. The corresponding algorithm is adopted in applications like movie ranking and driver ranking, and have been shown to be very effective.

4.3.3   Ranking Using Cumulative Distribution Networks

In [16], a ranking method based on cumulative distribution networks is proposed, which can be regarded as an extension of ListNet and ListMLE.

As pointed out in [16], unlike standard regression or classification in which we predict outputs independently, in ranking we are interested in predicting structured outputs so that mis-ranking one object can significantly affect whether we correctly rank the other objects. For this purpose, cumulative distribution network (CDN) is used, which is an undirected graphical model whose joint cumulative density function (CDF) $F(z)$ over a set of random variables is represented as a product over functions defined over subsets of these variables. Mathematically,

$$F(z) = \prod_c \phi_c(z_c), \qquad (4.26)$$

where ϕ_c is a potential function, and c is a clique in the graphical model.

Fig. 4.2 The CDN model

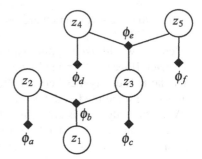

An example is shown in Fig. 4.2. In the example, the joint CDF
$F(z_1, z_2, z_3, z_4, z_5) = \phi_a(z_2)\phi_b(z_1, z_2, z_3)\phi_c(z_3)\phi_e(z_3, z_4, z_5)\phi_f(z_5)$.

When applied to learning to rank, the vertices in the CDN represent the documents to be ranked and the edges represent their preference relationships. Given each edge in the CDN, the likelihood of the preference represented by the edge can be defined by the $F(\cdot)$ function, which takes the following form in [16]:

$$F(z) = \prod_{e,e'} \frac{1}{1 + \exp(-\theta_1(f(x_u) - f(x_v))) + \exp(-\theta_2(f(x'_u) - f(x'_v)))}, \quad (4.27)$$

where e is the edge between nodes x_u and x_v and e' is the edge between nodes x'_u and x'_v.

Then, the negative log likelihood of all the preferences captured by the graph is used as the loss function for ranking. The algorithm that learns the ranking model by minimizing this loss function is called StructRank.

As discussed in [16], ListNet [4] and ListMLE [31] can be regarded as special cases of StructRank. Specifically, in these algorithms, the CDN function is defined with $m - 1$ multivariate sigmoids: the first sigmoid is concerned with $m - 1$ edges, the second with $m - 2$ edges, ..., the last with only one edge. Furthermore, in certain conditions, RankNet [3] can also be regarded as a special case of StructRank.

4.3.4 BoltzRank

In [30], a flexible ranking model is proposed, referred to as BoltzRank. BoltzRank utilizes a scoring function composed of individual and pairwise potentials to define a conditional probability distribution, in the form of a Boltzmann distribution, over all permutations of documents retrieved for a given query.

Specifically, given a set of documents $\mathbf{x} = \{x_j\}_{j=1}^m$, their ranking scores $\mathbf{s} = \{f(x_j)\}_{j=1}^m$, and the corresponding ranking π, the conditional energy of π given $\mathbf{s}$ is defined as follows:

$$E[\pi|\mathbf{s}] = \frac{2}{m(m-1)} \sum_{u<v} g_q(v-u) \cdot (s_{\pi^{-1}(u)} - s_{\pi^{-1}(v)}), \quad (4.28)$$

where g_q is any sign-preserving function, such as $g_q(x) = \alpha_q x$, and α_q is a query-dependent positive constant.

It is clear that $E[\pi|\mathbf{s}]$ represents the lack of compatibility between the relative document orderings given by π and those given by $\mathbf{s}$, with more positive energy indicating less compatibility.

Using the energy function, we can now define the conditional Boltzmann distribution over document permutations by exponentiating and normalizing, as follows:

$$P(\pi|\mathbf{s}) = \frac{1}{Z}\exp(-E[\pi|\mathbf{s}]), \quad Z = \sum_{\pi}\exp(-E[\pi|\mathbf{s}]). \qquad (4.29)$$

Note that it is costly to compute $P(\pi|\mathbf{s})$ and Z exactly since both of them contain sums over exponentially many permutations. In practice, however, efficient approximations of these quantities allow inference and learning in the model.

After defining the permutation probability as above, one can follow the idea in ListMLE [31] to define the loss function as the negative log likelihood of the ground truth permutation, or follow the idea in ListNet [4] to define the loss function as the K-L divergence between the permutation probability distribution given by the scoring function f and that given by the ground truth labels.

In addition, it is also possible to extend the above idea to optimize evaluation measures [30]. For example, suppose the evaluation measure is NDCG, then its expected value with respect to all the possible permutations can be computed as follows:

$$E[\text{NDCG}] = \sum_{\pi} P(\pi|\mathbf{s})\text{NDCG}(\pi, \mathbf{y}). \qquad (4.30)$$

Since only $P(\pi|\mathbf{s})$ in the above formula is a function of the ranking model f, and it is continuous and differentiable, one can use simple gradient based methods to effectively optimize (4.30). In order to avoid over fitting, one can also use the linear combination of the expected NDCG and the K-L divergence loss to guide the optimization process. Experimental results have shown that in this way, the BoltzRank method can outperform many other ranking methods, such as AdaRank [32] and ListNet [4].

4.4 Summary

As shown in this chapter, different kinds of listwise ranking algorithms have been proposed. Intuitively speaking, they model the ranking problem in a more natural way than the pointwise and pairwise approaches, and thus can address some problems that these two approaches have encountered. As we have discussed in the previous chapters, for the pointwise and pairwise approaches, the positional information is invisible to their loss functions, and they ignore the fact that some documents (or document pairs) are associated with the same query. Comparatively speaking, the listwise approach takes the entire set of documents associated with a query as the input instance and their ranked list (or their relevance degrees) as the output. In this way, it has the potential to distinguish documents from different queries, and to consider the positional information in the output ranked list (although not all listwise ranking methods have fully utilized this information) in its learning process. According to some previous studies [31], the performances of the listwise ranking algorithms are in general better than the pointwise or pairwise ranking algorithms. This is also verified by the discussions in Chap. 11, which is about the empirical ranking performances of different learning-to-rank algorithms, with the LETOR benchmark datasets as the experimental platform.

On the other hand, the listwise approach also has certain aspects to improve. For example, the training complexities of some listwise ranking algorithms (e.g., ListNet and BoltzRank) are high since the evaluation of their loss functions are permutation based. A more efficient learning algorithm is needed to make the listwise approach more practical. Moreover, the positional information has not been fully utilized in some listwise ranking algorithms, although it is visible to their loss functions. For example, there is no explicit position discount considered in the loss functions of ListNet and ListMLE. By introducing certain position discount factors, performance improvement of these algorithms can be expected.

4.5 Exercises

4.1 Derive the gradient of SoftNDCG with respect to the parameter w of the linear ranking model.

4.2 Actually SoftRank and Approximate Rank have strong connections with each other. In other words, SoftRank also leads to a kind of sigmoid approximation of the rank position. Prove this connection and show what kind of sigmoid function SoftRank uses implicitly.

4.3 Explain why the strategy used in SVM^{map} can find the most violated constraint, and analyze its time complexity.

4.4 Prove that under certain conditions the loss functions in ListNet and ListMLE are convex.

4.5 Suppose there are two features for learning to rank, please use a synthetic example to show the surface of NDCG in this case. Please further plot the surface of SoftNDCG and AppNDCG, and study their approximation to NDCG.

4.6 Please list the permutation probability models that can be used for ranking, except the Plackett–Luce model which has been well introduced in this chapter.

4.7 Show the possible ways of introducing position-based weights to ListMLE and ListNet, and compare their performances using the LETOR benchmark datasets.

4.8 Suppose the true loss for ranking is the permutation-level 0–1 loss. Prove that under certain assumptions on the permutation probability space, the optimal ranker given by ListMLE can also minimize the permutation-level 0–1 loss.

4.9 In practice, people care more about the correct ranking at the top positions of the ranking result. Accordingly, the true loss should not be a permutation-level 0–1 loss, but should be defined with respect to the top-k subgroup. Prove that in this new situation, ListMLE cannot lead to the optimal ranker in terms of the true loss. Show how to modify the loss function of ListMLE such that its minimization can minimize the top-k true loss.

References

1. Almeida, H., Goncalves, M., Cristo, M., Calado, P.: A combined component approach for finding collection-adapted ranking functions based on genetic programming. In: Proceedings of the 30th Annual International ACM SIGIR Conference on Research and Development in Information Retrieval (SIGIR 2007), pp. 399–406 (2007)
2. Andrews, S., Tsochantaridis, I., Hofmann, T.: Support vector machines for multiple-instance learning. In: Advances in Neural Information Processing Systems 15 (NIPS 2002), pp. 561–568 (2003)
3. Burges, C.J., Shaked, T., Renshaw, E., Lazier, A., Deeds, M., Hamilton, N., Hullender, G.: Learning to rank using gradient descent. In: Proceedings of the 22nd International Conference on Machine Learning (ICML 2005), pp. 89–96 (2005)
4. Cao, Z., Qin, T., Liu, T.Y., Tsai, M.F., Li, H.: Learning to rank: from pairwise approach to listwise approach. In: Proceedings of the 24th International Conference on Machine Learning (ICML 2007), pp. 129–136 (2007)
5. Chakrabarti, S., Khanna, R., Sawant, U., Bhattacharyya, C.: Structured learning for non-smooth ranking losses. In: Proceedings of the 14th ACM SIGKDD International Conference on Knowledge Discovery and Data Mining (KDD 2008), pp. 88–96 (2008)
6. Chapelle, O., Le, Q., Smola, A.: Large margin optimization of ranking measures. In: NIPS 2007 Workshop on Machine Learning for Web Search (2007)
7. Chapelle, O., Wu, M.: Gradient descent optimization of smoothed information retrieval metrics. Information Retrieval Journal. Special Issue on Learning to Rank 13(3), doi:10.1007/s10791-009-9110-3 (2010)
8. Fan, W., Fox, E.A., Pathak, P., Wu, H.: The effects of fitness functions on genetic programming based ranking discovery for web search. Journal of American Society for Information Science and Technology 55(7), 628–636 (2004)
9. Fan, W., Gordon, M., Pathak, P.: Discovery of context-specific ranking functions for effective information retrieval using genetic programming. IEEE Transactions on Knowledge and Data Engineering 16(4), 523–527 (2004)
10. Fan, W., Gordon, M., Pathak, P.: A generic ranking function discovery framework by genetic programming for information retrieval. Information Processing and Management 40(4), 587–602 (2004)
11. Fan, W., Gordon, M., Pathak, P.: Genetic programming-based discovery of ranking functions for effective web search. Journal of Management of Information Systems 21(4), 37–56 (2005)
12. Fan, W., Gordon, M., Pathak, P.: On linear mixture of expert approaches to information retrieval. Decision Support System 42(2), 975–987 (2006)
13. Fan, W., Gordon, M.D., Xi, W., Fox, E.A.: Ranking function optimization for effective web search by genetic programming: an empirical study. In: Proceedings of the 37th Hawaii International Conference on System Sciences (HICSS 2004), p. 40105 (2004)
14. Guiver, J., Snelson, E.: Learning to rank with softrank and Gaussian processes. In: Proceedings of the 31st Annual International ACM SIGIR Conference on Research and Development in Information Retrieval (SIGIR 2008), pp. 259–266 (2008)
15. Guiver, J., Snelson, E.: Bayesian inference for Plackett–Luce ranking models. In: Proceedings of the 26th International Conference on Machine Learning (ICML 2009), pp. 377–384 (2009)

16. Huang, J., Frey, B.: Structured ranking learning using cumulative distribution networks. In: Advances in Neural Information Processing Systems 21 (NIPS 2009) (2008)
17. Joachims, T.: A support vector method for multivariate performance measures. In: Proceedings of the 22nd International Conference on Machine Learning (ICML 2005), pp. 377–384 (2005)
18. Kass, R.E., Steffey, D.: Approximate Bayesian inference in conditionally independent hierarchical models. Journal of the American Statistical Association **84**(407), 717–726 (1989)
19. Le, Q., Smola, A.: Direct optimization of ranking measures. Tech. rep., arXiv:0704.3359 (2007)
20. Luce, R.D.: Individual Choice Behavior. Wiley, New York (1959)
21. Mallows, C.L.: Non-null ranking models. Biometrika **44**, 114–130 (1975)
22. McCullagh, P., Nelder, J.A.: Generalized Linear Models, 2nd edn. CRC Press, Boca Raton (1990)
23. Metzler, D.A., Croft, W.B., McCallum, A.: Direct maximization of rank-based metrics for information retrieval. Tech. rep., CIIR (2005)
24. Plackett, R.L.: The analysis of permutations. Applied Statistics **24**(2), 193–202 (1975)
25. Qin, T., Liu, T.Y., Li, H.: A general approximation framework for direct optimization of information retrieval measures. Information Retrieval **13**(4), 375–397 (2009)
26. Robertson, S., Zaragoza, H.: On rank-based effectiveness measures and optimization. Information Retrieval **10**(3), 321–339 (2007)
27. Talyor, M., Guiver, J., et al.: Softrank: optimising non-smooth rank metrics. In: Proceedings of the 1st International Conference on Web Search and Web Data Mining (WSDM 2008), pp. 77–86 (2008)
28. Trotman, A.: Learning to rank. Information Retrieval **8**(3), 359–381 (2005)
29. Tsochantaridis, I., Hofmann, T., Joachims, T., Altun, Y.: Support vector machine learning for interdependent and structured output space. In: Proceedings of the 21st International Conference on Machine Learning (ICML 2004), pp. 104–111 (2004)
30. Volkovs, M.N., Zemel, R.S.: Boltzrank: learning to maximize expected ranking gain. In: Proceedings of the 26th International Conference on Machine Learning (ICML 2009), pp. 1089–1096 (2009)
31. Xia, F., Liu, T.Y., Wang, J., Zhang, W., Li, H.: Listwise approach to learning to rank—theorem and algorithm. In: Proceedings of the 25th International Conference on Machine Learning (ICML 2008), pp. 1192–1199 (2008)
32. Xu, J., Li, H.: Adarank: a boosting algorithm for information retrieval. In: Proceedings of the 30th Annual International ACM SIGIR Conference on Research and Development in Information Retrieval (SIGIR 2007), pp. 391–398 (2007)
33. Xu, J., Liu, T.Y., Lu, M., Li, H., Ma, W.Y.: Directly optimizing IR evaluation measures in learning to rank. In: Proceedings of the 31st Annual International ACM SIGIR Conference on Research and Development in Information Retrieval (SIGIR 2008), pp. 107–114 (2008)
34. Yeh, J.Y., Lin, J.Y., et al.: Learning to rank for information retrieval using genetic programming. In: SIGIR 2007 Workshop on Learning to Rank for Information Retrieval (LR4IR 2007) (2007)
35. Yilmaz, E., Robertson, S.: On the choice of effectiveness measures for learning to rank. Information Retrieval Journal. Special Issue on Learning to Rank 13(3), doi:10.1007/s10791-009-9116-x (2010)
36. Yue, Y., Finley, T., Radlinski, F., Joachims, T.: A support vector method for optimizing average precision. In: Proceedings of the 30th Annual International ACM SIGIR Conference on Research and Development in Information Retrieval (SIGIR 2007), pp. 271–278 (2007)
37. Zoeter, O., Taylor, M., Snelson, E., Guiver, J., Craswell, N., Szummer, M.: A decision theoretic framework for ranking using implicit feedback. In: SIGIR 2008 Workshop on Learning to Rank for Information Retrieval (LR4IR 2008) (2008)

Chapter 5
Analysis of the Approaches

Abstract In this chapter, we introduce the analysis on the major approaches to learning to rank, which looks at the relationship between their loss functions and widely used evaluation measures. Basically, for many state-of-the-art ranking methods, it has been proven that their loss functions can upper bound measure-based ranking errors. Therefore the minimization of these loss functions can lead to the optimization of the evaluation measures in certain situations.

5.1 Overview

In the previous three chapters, we have introduced the pointwise, pairwise, and listwise approaches to learning to rank. The major differences between these approaches lie in their loss functions. Note that it is these loss functions that guide the learning process; however, the evaluation of the learned ranking models is based on the evaluation measures. Therefore, an important issue to discuss is the relationship between the loss functions used in these approaches and the evaluation measures. This is exactly the motivation of this chapter.[1]

Without loss of generality, we will take NDCG (more accurately NDCG@m, where m is the total number of documents associated with the query) and MAP as examples in the discussions. Furthermore, we assume that the ground-truth label is given as the relevance degree of each document.

5.2 The Pointwise Approach

As mentioned in Chap. 2, Cossock and Zhang [4] have established the theoretical foundation for reducing ranking to regression.[2] Given $\mathbf{x} = \{x_j\}_{j=1}^m$, a set of docu-

---

[1]Note that based on our current understanding on the issue, it is possible that we cannot establish connections between the evaluation measures and the loss functions in all the algorithms introduced in the previous chapters. However, it would be already very helpful if we can establish the connection between some popularly used loss functions and evaluation measures.

[2]Note that the bounds given in the original papers are with respect to DCG, and here we give their equivalent form in terms of NDCG for ease of comparison.

T.-Y. Liu, *Learning to Rank for Information Retrieval*,
DOI 10.1007/978-3-642-14267-3_5, © Springer-Verlag Berlin Heidelberg 2011

ments associated with training query q, and the ground truth $\mathbf{y} = \{y_j\}_{j=1}^m$ of these documents in terms of multiple ordered categories, suppose a scoring function f is used to rank these documents and the ranked list is π. A theory has been proven that the NDCG-based ranking error can be bounded by the following regression loss:

$$1 - \text{NDCG}(\pi, \mathbf{y}) \leq \frac{1}{Z_m} \left(2 \sum_{j=1}^m \eta(j)^2 \right)^{\frac{1}{2}} \left(\sum_{j=1}^m (f(x_j) - y_j)^2 \right)^{\frac{1}{2}}, \quad (5.1)$$

where Z_m is the maximum DCG value and $\eta(j)$ is the discount factor used in NDCG.

In other words, if one can really minimize the regression loss to zero, one can also minimize $(1 - \text{NDCG})$ to zero. This seems to be a very nice property of the regression-based methods.

With similar proof techniques to those used in [4], Li et al. [7] show that $(1 - \text{NDCG})$ can also be bounded by the multi-class classification loss as shown below (it is assumed that there are five ordered categories, i.e., $K = 5$ in the inequality):

$$1 - \text{NDCG}(\pi, \mathbf{y}) \leq \frac{15}{Z_m} \sqrt{2 \left(\sum_{j=1}^m \eta(j)^2 - m \prod_{j=1}^m \eta(j)^{\frac{2}{m}} \right)} \cdot \sqrt{\sum_{j=1}^m I_{\{y_j \neq \hat{y}_j\}}}, \quad (5.2)$$

where $\hat{y}_j$ is the prediction on the label of x_j by the multi-class classifier, and $f(x_j) = \sum_{k=0}^{K-1} k \cdot P(\hat{y}_j = k)$.

In other words, if one can really minimize the classification loss to zero, one can also minimize $(1 - \text{NDCG})$ to zero at the same time.

However, on the other hand, please note that when $(1 - \text{NDCG})$ is zero (i.e., the documents are perfectly ranked), the regression loss and the classification loss might not be zero (and can still be very large). That is, the minimization of the regression loss and the classification loss is only a sufficient condition but not a necessary condition for optimal ranking in terms of NDCG.

Let us have a close look at the classification bound in (5.2) with an example.[3] Note that a similar example has been given in [1] to show the problem of reducing ranking to binary classification.

Suppose for a particular query q, we have four documents (i.e., $m = 4$) in total, and their ground-truth labels are 4, 3, 2, and 1, respectively (i.e., $y_1 = 4$, $y_2 = 3$, $y_3 = 2$, $y_4 = 1$). We use the same discount factor and gain function as in [7]. Then it is easy to compute that $Z_m = \sum_{j=1}^4 \frac{1}{\log(j+1)} (2^{y_j - 1}) \approx 21.35$.

Then, suppose the outputs of the multi-class classifier are $\hat{y}_1 = 3$, $\hat{y}_2 = 2$, $\hat{y}_3 = 1$, and $\hat{y}_4 = 0$, with 100% confidence in the prediction for each document. It is easy to compute that $(1 - \text{NDCG})$ is 0 and we actually get a perfect ranking based on the classifier. However, in terms of multi-class classification, we made errors in all four

---

[3]One can get similar results for the regression bound given in inequality (5.1).

documents, i.e., $\sum_{j=1}^{m} I_{\{y_j \neq \hat{y}_j\}} = 4$. Furthermore, if we compute the bound given by (5.2), we obtain

$$\frac{15}{Z_m} \sqrt{2 \left(\sum_{j=1}^{m} \left(\frac{1}{\log(j+1)} \right)^2 - m \prod_{j=1}^{m} \left(\frac{1}{\log(j+1)} \right)^{\frac{2}{m}} \right) \cdot \sum_{j=1}^{m} I_{\{y_j \neq \hat{y}_j\}}}$$

$$\approx \frac{24.49}{21.35} = 1.15.$$

It is clear that the bound is meaningless since it is larger than one. Actually the loose bound is not difficult to understand. The left-hand side of inequality (5.2) contains positional information, while the right-hand side does not. When the same amount of classification loss occurs in different positions, the ranking error will be quite different. In order to make the inequality always hold, the price one has to pay is that the bound must be very loose.

Note that there are no results yet in the literature to connect the aforementioned regression loss and classification loss to the MAP-based ranking error.

5.3 The Pairwise Approach

In [6], the following theoretical result has been obtained, in addition to the design of the Ranking SVM algorithm.

$$1 - \text{MAP}(\pi, \mathbf{y}) \leq 1 - \frac{1}{m_1} \left(L^{\text{pairwise}}(f; \mathbf{x}, \mathbf{y}) + C_{m_1+1}^2 \right)^{-1} \left(\sum_{j=1}^{m_1} \sqrt{j} \right)^2, \quad (5.3)$$

where

$$L^{\text{pairwise}}(f; \mathbf{x}, \mathbf{y}) = \sum_{t=1}^{m-1} \sum_{i=1, y_i < y_t}^{m} \phi \big(f(x_t) - f(x_i) \big), \quad (5.4)$$

which corresponds to the query-level sum of the pairwise loss function in Ranking SVM, RankBoost, and RankNet, when the ϕ function is the hinge function ($\phi(z) = (1-z)_+$), exponential function ($\phi(z) = e^{-z}$), and logistic function ($\phi(z) = \log(1 + e^{-z})$), respectively.

The above result basically says that there is a connection between $(1 - \text{MAP})$ and the loss functions in several pairwise ranking algorithms. When minimizing the pairwise loss functions, $(1 - \text{MAP})$ can also be minimized. However, this result has the following limitations.

- The connection revealed in the above result is somehow indirect. It is not clear whether the pairwise loss functions and $(1 - \text{MAP})$ have a direct relationship, e.g., the pairwise loss functions are upper bounds of $(1 - \text{MAP})$.

- According to the above result, the zero values of the pairwise loss functions are not sufficient conditions for the zero value of $(1 - \text{MAP})$. This is a weaker condition than our intuition. As we know, when the pairwise loss function takes a zero value, every document pair is correctly classified, and correspondingly we will get the correct ranking of the documents. In this case, we should find that $(1 - \text{MAP})$ equals zero.
- It is not clear whether a similar result can also be obtained for NDCG and other evaluation measures.

The authors of [3] try to overcome the aforementioned limitations. Specifically, they show that many of the pairwise loss functions are upper bounds of a quantity, named the essential loss for ranking. Furthermore, the essential loss is an upper bound of $(1 - \text{NDCG})$ and $(1 - \text{MAP})$, and therefore these loss functions are also upper bounds of $(1 - \text{NDCG})$ and $(1 - \text{MAP})$.

To better illustrate this result, we first introduce the concept of the essential loss, which is constructed by modeling ranking as a sequence of classification tasks.

Given a set of documents $\mathbf{x}$ and their ground-truth permutation $\pi_y \in \Omega_y$, the ranking problem can be decomposed into several sequential steps. For each step t, one tries to distinguish $\pi_y^{-1}(t)$, the document ranked at the tth position in π_y, from all the documents ranked below the tth position in π_y, using a scoring function f. If we denote $\mathbf{x}_{(t)} = \{x_{\pi_y^{-1}(t)}, \ldots, x_{\pi_y^{-1}(m)}\}$, one can define a classifier based on f, whose target output is $\pi_y^{-1}(t)$,

$$T \circ f(\mathbf{x}_{(t)}) = \arg \max_{j \in \{\pi_y^{-1}(t), \ldots, \pi_y^{-1}(m)\}} f(x_j). \tag{5.5}$$

It is clear that there are $m - t$ possible outputs of this classifier, i.e., $\{\pi_y^{-1}(t), \ldots, \pi_y^{-1}(m)\}$. The 0–1 loss for this classification task can be written as follows, where the second equation is based on the definition of $T \circ f$:

$$L_t\left(f; \mathbf{x}(t), \pi_y^{-1}(t)\right) = I_{\{T \circ f(\mathbf{x}_{(t)}) \neq \pi_y^{-1}(t)\}}$$

$$= 1 - \prod_{j=t+1}^{m} I_{\{f(\pi_y^{-1}(t)) > f(\pi_y^{-1}(j))\}}. \tag{5.6}$$

A simple example is given in Fig. 5.1 to illustrate the aforementioned process of decomposition. Suppose there are three objects, A, B, and C, and the corresponding ground-truth permutation is $\pi_y = (A, B, C)$. Suppose the output of the scoring function for these objects is $(2, 3, 1)$, and accordingly the predicted ranked list is $\pi = (B, A, C)$. At step one of the decomposition, the scoring function predicts object B to be at the top of the list. However, A should be on the top according to π_y. Therefore, a prediction error occurs. For step two, we remove A from both π_y and π. Then the scoring function predicts object B to be on the top of the remaining list. This is in accordance with π_y and there is no prediction error. After that, we further remove object B, and it is easy to verify there is no prediction error in step three

$$\begin{matrix} \pi_y \\ \begin{pmatrix} A \\ B \\ C \end{pmatrix} \end{matrix} \quad \begin{matrix} \pi \\ \begin{pmatrix} B \\ A \\ C \end{pmatrix} \end{matrix} \quad \begin{matrix} \text{incorrect} \\ \Longrightarrow \\ \text{remove } A \end{matrix} \quad \begin{matrix} \pi_y \\ \begin{pmatrix} B \\ C \end{pmatrix} \end{matrix} \quad \begin{matrix} \pi \\ \begin{pmatrix} B \\ C \end{pmatrix} \end{matrix} \quad \begin{matrix} \text{correct} \\ \Longrightarrow \\ \text{remove } B \end{matrix} \quad \begin{matrix} \pi_y \\ (C) \end{matrix} \quad \begin{matrix} \pi \\ (C) \end{matrix}$$

Fig. 5.1 Modeling ranking as a sequence of classifications

either. Overall speaking, the scoring function makes one error in this sequence of classification tasks.

By further assigning a non-negative weight $\beta(t)$ $(t = 1, \ldots, m - 1)$ to the classification task at the tth step, which represents its importance to the entire sequence, we will get the weighted sum of the classification errors of all individual tasks,

$$L_\beta(f; \mathbf{x}, \pi_y) \triangleq \sum_{t=1}^{m-1} \beta(t) \left(1 - \prod_{i=t+1}^{m} I_{\{f(x_{\pi^{-1}(t)}) > f(x_{\pi^{-1}(i)})\}} \right). \qquad (5.7)$$

Then we define the minimum value of the weighted sum over all the permutations in Ω_y as the *essential loss for ranking*.

$$L_\beta(f; \mathbf{x}, \Omega_y) = \min_{\pi_y \in \Omega_y} L_\beta(f; \mathbf{x}, \pi_y). \qquad (5.8)$$

It has been proven in [3] that the essential loss is an upper bound of both $(1 - \text{NDCG})$ and $(1 - \text{MAP})$, as shown in the following theorem. As a result, the minimization of it will lead to the effective maximization of NDCG and MAP.

Theorem 5.1 *Suppose there are m_k documents with rating k and $\sum_{i=k^*}^{K} m_i > 0$ (when computing MAP, labels smaller than k^* are regarded as irrelevant), then $\forall f$, the following inequalities hold,*

(1) $1 - \text{NDCG}(\pi, \mathbf{y}) \le \dfrac{1}{Z_m} L_{\beta_1}(f; \mathbf{x}, \Omega_y),$

   where $\beta_1(t) = G(l_{\pi_y^{-1}(t)}) \eta(t), \forall \pi_y \in \Omega_y;$

(2) $1 - \text{MAP}(\pi, \mathbf{y}) \le \dfrac{1}{\sum_{i=k^*}^{K} m_i} L_{\beta_2}(f; \mathbf{x}, \Omega_y),$ *where $\beta_2(t) \equiv 1$.*

As compared to the bounds for the pointwise approach as given in the previous section, one can see that the essential loss has a nicer property. When $(1 - \text{NDCG})$ or $(1 - \text{MAP})$ is zero, the essential loss is also zero. In other words, the zero value of the essential loss is not only a sufficient condition but also a necessary condition of the zero values of $(1 - \text{NDCG})$ and $(1 - \text{MAP})$.

Furthermore, it has been proven in [3] that the essential loss is the lower bound for many pairwise loss functions.

Theorem 5.2 *The pairwise losses in Ranking SVM, RankBoost, and RankNet are all upper bounds of the essential loss, i.e.,*

$$L_\beta(f; \mathbf{x}, \Omega_y) \le \left(\max_{1 \le t \le m-1} \beta(t) \right) L^{\text{pairwise}}(f; \mathbf{x}, \mathbf{y}),$$

where $L^{\text{pairwise}}(f; \mathbf{x}, \mathbf{y})$ is defined as in (5.4).

Therefore, the minimization of the loss functions in the aforementioned pairwise ranking algorithms will all lead to the minimization of the essential loss. Further considering the relationship between the essential loss and $(1 - \text{NDCG})$ as well as $(1 - \text{MAP})$, we have

$$1 - \text{NDCG}(\pi, \mathbf{y}) \le \frac{G(K-1)\eta(1)}{Z_m} L^{\text{pairwise}}(f; \mathbf{x}, \mathbf{y}); \qquad (5.9)$$

$$1 - \text{MAP}(\pi, \mathbf{y}) \le \frac{1}{\sum_{i=k^*}^{K} m_i} L^{\text{pairwise}}(f; \mathbf{x}, \mathbf{y}). \qquad (5.10)$$

In other words, optimizing these pairwise loss functions can minimize $(1 - \text{NDCG})$ and $(1 - \text{MAP})$.

5.4 The Listwise Approach

5.4.1 Non-measure-Specific Loss

We take ListMLE as an example of the listwise ranking algorithms in this sub-category. The following results have been proven in [3], which characterizes the relationship between the likelihood loss and $(1 - \text{NDCG})$ as well as $(1 - \text{MAP})$.

Theorem 5.3 *The loss function in ListMLE is an upper bound of the essential loss, i.e.,*

$$L_\beta(f; \mathbf{x}, \Omega_y) \le \frac{1}{\ln 2} \left(\max_{1 \le t \le m-1} \beta(t) \right) L^{\text{ListMLE}}(f; \mathbf{x}, \Omega_y). \qquad (5.11)$$

Further considering the relationship between the essential loss and $(1 - \text{NDCG})$ as well as $(1 - \text{MAP})$, one can come to the following conclusions.

$$1 - \text{NDCG}(\pi, \mathbf{y}) \le \frac{G(K-1)\eta(1)}{Z_m \ln 2} L^{\text{ListMLE}}(f; \mathbf{x}, \Omega_y); \qquad (5.12)$$

$$1 - \text{MAP}(\pi, \mathbf{y}) \le \frac{1}{\ln 2 \sum_{i=k^*}^{K} m_i} L^{\text{ListMLE}}(f; \mathbf{x}, \Omega_y). \qquad (5.13)$$

Note that the proof of the above theorem highly depends on the form of the likelihood loss. Extensive work is needed to generalize this result to other listwise ranking methods.

5.4.2 Measure-Specific Loss

The other sub-category of the listwise approach optimizes a measure-specific loss function. Therefore, the discussion on the relationship between such a loss function and the corresponding evaluation measure will be more straightforward, since they have natural connections. For example, some algorithms are explicitly designed as optimizing an upper bound of the measure-based ranking error. However, also because of this, one may not be satisfied with the relationship of an "upper bound" and want to know more about these algorithms. For this purpose, a new quantity named "tendency correlation" is introduced in [5]. Basically, the tendency correlation measures the relationship between a surrogate measure (e.g., SoftNDCG and AppNDCG) and the corresponding evaluation measure. Note that in the definition, for simplicity, we assume the use of a linear scoring function, i.e., $f = w^T x$.

Definition 5.1 (ε Tendency Correlation) Given an evaluation measure M and a surrogate measure $\tilde{M}$, the documents $\mathbf{x}$ associated with a query, and their ground-truth labels $\mathbf{y}$, for two ranking models w_1 and w_2, denote

$$\varepsilon(\beta, w_1, w_2, \mathbf{x}, \mathbf{y}) = \left| \beta \left[\tilde{M}(w_1; \mathbf{x}, \mathbf{y}) - \tilde{M}(w_2; \mathbf{x}, \mathbf{y}) \right] \right.$$
$$\left. - \left[M(w_1; \mathbf{x}, \mathbf{y}) - M(w_2; \mathbf{x}, \mathbf{y}) \right] \right|, \quad \beta \geq 0.$$

Then the tendency correlation between $\tilde{M}$ and M is defined as

$$\varepsilon = \min_{\beta} \max_{w_1, w_2, \mathbf{x}, \mathbf{y}} \varepsilon(\beta, w_1, w_2, \mathbf{x}, \mathbf{y}),$$

and we say $\tilde{M}$ has ε tendency correlation with M.

When ε is small, we say that the surrogate measure has a strong tendency correlation with the evaluation measure. According to [5], the tendency correlation as defined above has the following properties.

1. The tendency correlation is invariant to the linear transformation of the surrogate measure. In other words, suppose a surrogate measure has ε tendency correlation with the evaluation measure, and we transform the surrogate measure by means of translation and/or scaling, the new measure we obtain will also have ε tendency correlation with the evaluation measure.
2. If ε is zero, the surrogate measure will have the same shape as the evaluation measure (i.e., their only differences lie in a linear transformation). If ε is very small, the surrogate measure will have a very similar shape to the evaluation measure.
3. It has been theoretically justified that when ε approaches zero, the optimum of the surrogate measure will converge to the optimum of the evaluation measure [5].

One may have noticed that the definition of tendency correlation reflects the worst-case disagreement between an evaluation measure and a surrogate measure.

The advantages of using the worst-case disagreement as compared to using other kinds of disagreements, such as the average disagreement, are as follows. First, the worst-case disagreement is much easier to calculate because it does not consider the detailed data distribution. Second, if the worst-case disagreement is small, we can come to the conclusion that the two measures are highly correlated everywhere in the input space. As a result, their global optima will be very close. However, if we only know that the average disagreement is small, it is still possible that the two measures are largely different at some point. As a consequence, we can hardly come to any conclusion about the relationship between the global optima of the two measures.

With the concept of tendency correlation, the following theoretical results have been proven in [5]. Note that although not all the direct optimization methods introduced in this book have been covered by these theorems, the extensions to them are not difficult based on the proofing techniques given in [5]. The following results have also been empirically verified in [5] and are in accordance with the experimental results reported in other work.

For SoftRank [9], one can obtain the following result, which basically indicates that SoftNDCG can have an arbitrarily strong tendency correlation with NDCG if the parameter σ_s is set to be arbitrarily small.

Theorem 5.4 *For query q, suppose its associated documents and ground-truth labels are $(\mathbf{x}, \mathbf{y})$. Assume $\forall i$ and j, $|w^T x_i - w^T x_j| \geq \delta > 0$,[4] and $\forall q, m \leq m_0$. If $\sigma_s < \dfrac{\delta}{2\mathrm{erf}^{-1}(\sqrt{\frac{5m_0-9}{5m_0-5}})}$, then SoftNDCG has ε tendency correlation with NDCG and ε satisfies that*

$$\varepsilon \leq N \cdot 2^L \cdot (\varepsilon_1 + \varepsilon_2),$$

where $\varepsilon_1 = \dfrac{(m_0-1)\sigma_s}{2\delta\sqrt{\pi}} e^{-\frac{\delta^2}{4\sigma_s^2}}$, $\varepsilon_2 = \sqrt{\dfrac{\varepsilon_3}{1-5\varepsilon_3}} + 5\varepsilon_3$, $\varepsilon_3 = \dfrac{(m_0-1)[1-\mathrm{erf}^2(\frac{\delta}{2\sigma_s})]}{4}$.

For Approximate Rank [8], one can obtain the following result for AppNDCG, which indicates that AppNDCG can have an arbitrarily strong tendency correlation with NDCG if the parameter α is set to be arbitrarily large. As for AppMAP, similar result can also be obtained.

Theorem 5.5 *For query q, suppose its associated documents and ground-truth labels are $(\mathbf{x}, \mathbf{y})$. Assume $\forall i$ and j, $|w^T x_i - w^T x_j| \geq \delta > 0$, and $\forall q, m \leq m_0$. If $\alpha > \dfrac{\log(\max\{0, 2m_0-3\})}{\delta}$, then AppNDCG has ε tendency correlation with NDCG and*

---

[4]Note that this assumption is made in order that one can obtain a unique ranked list by sorting the scores assigned by $f = w^T x$ to the documents. Otherwise, the output of the ranking model will correspond to different evaluation-measure values depending upon how we rank the documents with the same score. In this case, it makes no sense how the evaluation measure is directly optimized since the evaluation measure is not a single value.

ε satisfies

$$\varepsilon < \frac{\frac{m_0 - 1}{\exp(\delta\alpha)} + 1}{\ln 2}.$$

For SVM^{map} [11], SVM^{ndcg} [2], and PermuRank [10], one can obtain the result as shown in the following theorem. Basically, one can always find an infinite number of samples on which the tendency correlations between their surrogate measures and the evaluation measures cannot be as strong as expected.

Theorem 5.6 *Let $\tilde{M}$ represent the surrogate measure used in SVM^{map} [11], SVM^{ndcg} [2], or PermuRank [10], and M the corresponding evaluation measure. Then for $\forall w_1 \neq 0$, and $\forall \Delta \in (0, 1)$, there exist $w_2 \neq 0$ and an infinite number of $(\mathbf{x}^*, \mathbf{y}^*)$ which satisfy*

$$-\text{sgn}\big[\tilde{M}(w_1; \mathbf{x}^*, \mathbf{y}^*) - \tilde{M}(w_2; \mathbf{x}^*, \mathbf{y}^*)\big]\big[M(w_1; \mathbf{x}^*, \mathbf{y}^*) - M(w_2; \mathbf{x}^*, \mathbf{y}^*)\big] \geq \Delta.$$

And therefore $\varepsilon \geq \Delta$.

According to the aforementioned theorems, we have the following discussions.

1. The surrogate measures optimized by SoftRank and Approximate Rank can have arbitrarily strong tendency correlation with the evaluation measures on any kind of data, when the parameters in these algorithms are appropriately set. Therefore, the evaluation measures can be effectively optimized by such methods and the corresponding ranking performance can be high.

2. The surrogate measures optimized by SVM^{map} [11], SVM^{ndcg} [2], and Permu-Rank [10] cannot have an arbitrarily strong tendency correlation with the evaluation measures on certain kinds of data. Note that the distribution of data is usually unknown in practical learning-to-rank setting. Thus, in practice, the ranking performance of these methods may not be very high, and may vary according to different datasets.

3. Considering that the evaluation measures usually contain many local optima with respect to the ranking model, a surrogate measure will also have many local optima when it has very strong tendency correlation with the evaluation measure. As a result, robust strategies that are able to find the global optimum are required when performing optimization. In other words, there is a trade-off between choosing a surrogate measure that has a strong tendency correlation with the evaluation measure and choosing a surrogate measure that can be easily optimized. This trade-off issue is especially critical for algorithms like SoftRank and Approximate Rank since they utilize the gradient descent techniques for optimization, which is easy to be trapped into a local optimum.

5.5 Summary

In this chapter, we have reviewed the relationships between different loss functions in learning to rank and the evaluation measures. The discussions well explain why different learning-to-rank algorithms perform reasonably well in practice (see the experimental results in Chap. 11).

While the analyses introduced in this chapter look quite nice, there are still several issues that have not been well solved. First, although the essential loss can be used to explain the relationship between measure-based ranking errors and several loss functions in the pairwise and listwise approaches, it is not clear whether it can be extended to explain the pointwise approach. Second, from a machine learning point of view, being an upper bound of the evaluation measure might not be sufficient for being a good loss function. The reason is that what we really care about is the optimal solution with regards to the loss function. Even if a loss can be the upper bound of a measure-based ranking error everywhere, its optimum may still be far away from the optimum of the measure-based ranking error.

The discussions on the "tendency correlation" are one step toward solving the second problem as aforementioned. However, the condition of the tendency correlation still seems too strong and sometimes not necessary. A more principled solution should be obtained by investigating the so-called "consistency" of the loss functions, which exactly describes whether the optima with regards to the loss function and the measure can be the same. Consistency of learning methods has been well studied in classification, but not yet for ranking. More discussions on this topic can be found in Part VI. Please note that the discussions on consistency are valid only when the number of training examples approaches infinity. However, in practice, the number of training examples is always limited. In this case, the theoretical analysis might not be always in accordance with experimental results [9]. Therefore, a lot of work needs to be further done in order to predict the performance of a learning-to-rank method when the sample size is small.

5.6 Exercises

5.1 In this chapter, the relationship between $(1 - MAP)$ and the pairwise loss functions as well as the listwise loss function have been discussed. However, there is nothing on whether the relationship between $(1 - MAP)$ and the pointwise loss function also exists. Please make discussions on this by extending the results in [4] and [7].

5.2 Investigate whether one can get similar results as presented in this chapter for other evaluation measures, such as MRR and RC.

5.3 Show that although the zero value of the essential loss is a necessary and sufficient condition for the zero value of $(1 - NDCG)$ and $(1 - MAP)$, when the value is very small but non-zero, $(1 - NDCG)$ and $(1 - MAP)$ will not necessarily be very small.

5.4 Study other applications of modeling ranking as a sequence of classification tasks, rather than defining the essential loss and deriving the upper bounds.

5.5 It has been observed in [9] that sometimes optimizing one measure on the training data might not lead to the best ranking performance on the test data in terms of the same measure. Explain this phenomenon and design an experiment to validate your hypothesis.

References

1. Agarwal, S., Graepel, T., Herbrich, R., Har-Peled, S., Roth, D.: Generalization bounds for the area under the roc curve. Journal of Machine Learning **6**, 393–425 (2005)
2. Chakrabarti, S., Khanna, R., Sawant, U., Bhattacharyya, C.: Structured learning for non-smooth ranking losses. In: Proceedings of the 14th ACM SIGKDD International Conference on Knowledge Discovery and Data Mining (KDD 2008), pp. 88–96 (2008)
3. Chen, W., Liu, T.-Y., Lan, Y., Ma, Z., Li, H.: Ranking measures and loss functions in learning to rank. In: Advances in Neural Information Processing Systems 22 (NIPS 2009), pp. 315–323 (2010)
4. Cossock, D., Zhang, T.: Subset ranking using regression. In: Proceedings of the 19th Annual Conference on Learning Theory (COLT 2006), pp. 605–619 (2006)
5. He, Y., Liu, T.-Y.: Tendency correlation analysis for direct optimization methods. Information Retrieval **13**(6), 657–688 (2010)
6. Joachims, T.: Optimizing search engines using clickthrough data. In: Proceedings of the 8th ACM SIGKDD International Conference on Knowledge Discovery and Data Mining (KDD 2002), pp. 133–142 (2002)
7. Li, P., Burges, C., Wu, Q.: McRank: learning to rank using multiple classification and gradient boosting. In: Advances in Neural Information Processing Systems 20 (NIPS 2007), pp. 845–852 (2008)
8. Qin, T., Liu, T.-Y., Li, H.: A general approximation framework for direct optimization of information retrieval measures. Information Retrieval **13**(4), 375–397 (2009)
9. Talyor, M., Guiver, J., et al.: Softrank: optimising non-smooth rank metrics. In: Proceedings of the 1st International Conference on Web Search and Web Data Mining (WSDM 2008), pp. 77–86 (2008)
10. Xu, J., Liu, T.-Y., Lu, M., Li, H., Ma, W.-Y.: Directly optimizing IR evaluation measures in learning to rank. In: Proceedings of the 31st Annual International ACM SIGIR Conference on Research and Development in Information Retrieval (SIGIR 2008), pp. 107–114 (2008)
11. Yue, Y., Finley, T., Radlinski, F., Joachims, T.: A support vector method for optimizing average precision. In: Proceedings of the 30th Annual International ACM SIGIR Conference on Research and Development in Information Retrieval (SIGIR 2007), pp. 271–278 (2007)

Part III
Advanced Topics in Learning to Rank

In this part, we will introduce some advanced learning-to-rank tasks, which are in a sense orthogonal to the three major approaches to learning to rank. On the one hand, these tasks consider specific issues that are not core issues in the major approaches. On the other hand, for any of the three major approaches, the technologies introduced in this part can be used to enhance them or extend them for new scenarios of learning to rank.

First, we will introduce the task of relational ranking, which explicitly explores the relationship between documents. This is actually an abstract of many real ranking scenarios. For example, in topic distillation, we will consider the parent–child hierarchy of webpages in the sitemap, and in diversity search, we would not like to see many topically similar documents ranked in the top positions. Here the parent–child hierarchy and topical similarity are both what we call "relationships".

Second, we will introduce the task of query-dependent ranking. As we can see, although different kinds of loss functions have been studied, it is usually assumed that a single ranking function is learned and used to answer all kinds of queries. Considering that queries might be very different in semantics and user intentions, it is highly preferred that the ranking function can incorporate query differences.

Third, we will introduce the task of semi-supervised ranking. As one may have noticed, most of the learning-to-rank work introduced in this book requires full supervision in order to learn the ranking model. However, it is always costly to get the fully labeled data, and it is much cheaper to obtain unlabeled queries and documents. It is therefore interesting and important how to leverage these unlabeled data to enhance learning to rank.

Last, we will introduce the task of transfer ranking. In practice, we will often encounter the situation that the training data for one ranking task are insufficient, but the task is related to another task with sufficient training data. For example, commercial search engines usually collect a large number of labeled data for their main-stream market, however, they cannot do the same thing for every niche market. Then how to leverage the training data from the main-stream market to make the ranking model for the niche market more effective and reliable is a typical transfer ranking problem.

After reading this part, the readers are expected to understand the advanced topics that we introduce, and to be able to formulate other advanced learning-to-rank tasks based on their understandings of various ranking problems in information retrieval.

Chapter 6
Relational Ranking

Abstract In this chapter, we introduce a novel task for learning to rank, which does not only consider the properties of each individual document in the ranking process, but also considers the inter-relationship between documents. According to different relationships (e.g., similarity, preference, and dissimilarity), the task may correspond to different real applications (e.g., pseudo relevance feedback, topic distillation, and search result diversification). Several approaches to solve this new task are reviewed in this chapter, and future research directions along this line are discussed.

As shown in previous chapters, in most cases it is assumed that the ranked list is generated by sorting the documents according to their scores outputted by a scoring function. That is, the hypothesis h can be represented as sort $\circ f(\mathbf{x})$, and function f works on a single document independent of other documents. However, in some practical cases, the relationship between documents should be considered, and only defining the scoring function f on single documents is not appropriate. For example, in the task of topic distillation [3], if a page and its parent page (in the sitemap) are similarly relevant to the query, it is desirable to rank the parent page above the child page. In the scenario of pseudo relevance feedback, it is assumed that documents that are similar in their content should be ranked close to each other even if their relevance features are different. In the scenario of search result diversification, it is not a good idea to rank documents that are topically very similar all on the top positions.

In the literature, there are several attempts on introducing inter-relationship between documents to the ranking process. For ease of reference, we call such kinds of ranking "relational ranking". In this chapter, we will first introduce a unified framework for relational ranking [6, 7], and then introduce several pieces of work that targets specifically at search result diversification [1, 4, 8, 9].

T.-Y. Liu, *Learning to Rank for Information Retrieval*,
DOI 10.1007/978-3-642-14267-3_6, © Springer-Verlag Berlin Heidelberg 2011

6.1 General Relational Ranking Framework

6.1.1 Relational Ranking SVM

By considering the relationship in the ranking process, the ranking function can be refined as follows, where the function h is named the relational ranking function [6],

$$h(\mathbf{x}) = h(f(\mathbf{x}), R). \tag{6.1}$$

Here f can still be understood as a scoring function operating on each single document, and the matrix R describes the relationship between documents. There are then several ways of parameterizing $h(\mathbf{x})$. For example, one can let f contain a parameter w, while the way of integrating f and R is pre-defined. One can also regard f as a fixed function, and h has its own parameter. The most complex case is that both f and h contain parameters. In [6], the first case is discussed, and it is assumed that

$$h(\mathbf{x}) = h(f(w, \mathbf{x}), R) = \arg\min_{\mathbf{z}}\{l_1(f(w, \mathbf{x}), \mathbf{z}) + \beta l_2(R, \mathbf{z})\}, \tag{6.2}$$

where l_1 is an objective function to guarantee that the ranking results given by h should not be too different from those given by f, and l_2 is another objective function to guarantee that the ranking results should be consistent with the relationship requirement R as much as possible.

Then, as for topic distillation and pseudo relevance feedback, the relational ranking function h is further realized and discussed. A nice thing is the existence of the closed-form solution of the relational ranking function in these two cases. For pseudo relevance feedback, one can eventually obtain,

$$h(f(w, \mathbf{x}), R) = (I + \beta(D - R))^{-1} f(w, \mathbf{x}), \tag{6.3}$$

where I denotes an identity matrix, and D is a diagonal matrix with $D_{i,i} = \sum_j R_{i,j}$.

For topic distillation, one has

$$h(f(w, \mathbf{x}), R) = (2I + \beta(2D - R - R^T))^{-1}(2f(w, \mathbf{x}) - \beta r), \tag{6.4}$$

where $r_j = \sum_i R_{i,k} - \sum_j R_{k,j}$.

One may have noticed that in the relational ranking framework, there are two different inputs: one is the features of the documents x, and the other is the relationship matrix R. This seems quite different from the setting of conventional learning to rank mentioned in previous chapters. However, if one notices that $x = \phi(q, d)$ and $R = \psi(d, d)$, one will realize that the inputs are still documents and queries. The only difference lies in that now we also care about the inter-dependency between documents, while previously each document was dealt with in an independent manner.

The relational ranking function can be substituted into any learning-to-rank algorithm introduced in this book for learning. In particular, the demonstration of learning the relational ranking function with Ranking SVM is given in [6]. The corresponding algorithm is named *"Relational Ranking SVM (RRSVM)"*. The matrix-form expressions of RRSVM are given as below.

RRSVM for pseudo relevance feedback can be represented as the following optimization problem:

$$\min_{w,\xi^{(i)}} \frac{1}{2}\|w\|^2 + \lambda \sum_{i=1}^{n} \mathbf{1}^{(i)^T} \xi^{(i)},$$

$$\text{s.t.} \quad C^{(i)} h\left(f\left(w, \mathbf{x}^{(i)}\right), R^{(i)}\right) \geq \mathbf{1}^{(i)} - \xi^{(i)}, \quad \xi^{(i)} \geq 0, \qquad (6.5)$$

$$h\left(f\left(w, \mathbf{x}^{(i)}\right), R^{(i)}\right) = \left(I - \beta\left(D^{(i)} - R^{(i)}\right)\right)^{-1} f\left(w, \mathbf{x}^{(i)}\right),$$

where $C^{(i)}$ denotes a constraint matrix for query q_i, $\mathbf{1}^{(i)}$ denotes a vector with all the elements being 1, and its dimension is the same as that of $\xi^{(i)}$. Each row of $C^{(i)}$ represents a pairwise constraint: one element is 1, one element is -1, and the other elements are all 0. For example, for query q_i, if the ground truth indicates that $y_1 > y_3$ and $y_2 > y_4$, then we have

$$C^{(i)} = \begin{pmatrix} 1 & 0 & -1 & 0 \\ 0 & 1 & 0 & -1 \end{pmatrix}.$$

RRSVM for topic distillation can be represented as the following optimization problem:

$$\min_{w,\xi^{(i)}} \frac{1}{2}\|w\|^2 + \lambda \sum_{i=1}^{n} \mathbf{1}^{(i)^T} \xi^{(i)},$$

$$\text{s.t.} \quad C^{(i)} h\left(f\left(w, \mathbf{x}^{(i)}\right), R^{(i)}\right) \geq \mathbf{1}^{(i)} - \xi^{(i)}, \quad \xi^{(i)} \geq 0,$$

$$h\left(f\left(w, \mathbf{x}^{(i)}\right), R^{(i)}\right) = \left(2I + \beta\left(2D^{(i)} - R^{(i)} - R^{(i)^T}\right)\right)^{-1} \qquad (6.6)$$

$$\times \left(2f\left(w, \mathbf{x}^{(i)}\right) - \beta r^{(i)}\right).$$

Further discussions indicate that although the ranking function becomes more complicated, one only needs to pre-process the features of all the documents associated with a given query, and can still use standard Ranking SVM toolkits to perform the learning and test.[1] Therefore the new algorithm is very feasible for practical use. Experimental results in [6] have shown that for particular learning tasks, RRSVM can significantly outperform Ranking SVM and other heuristic methods such as the pseudo relevance feedback method provided in the Lemur toolkit[2] and sitemap-based relevance propagation [5].

---

[1] For example, http://olivier.chapelle.cc/primal/ and http://www.cs.cornell.edu/People/tj/svm_light/svm_rank.html.

[2] http://www.lemurproject.org/.

Fig. 6.1 The C-CRF model

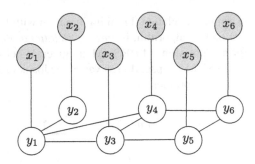

6.1.2 Continuous Conditional Random Fields

In [7], Qin et al. also investigate the formalization of relational ranking, but using a graphical model called continuous conditional random fields (C-CRF). C-CRF is a graphical model, as depicted in Fig. 6.1. In the conditioned undirected graph, a white vertex represents a ranking score, a gray vertex represents a document, an edge between two white vertexes represents the dependency between ranking scores, and an edge between a gray vertex and a white vertex represents the dependency of a ranking score on its document (content). (In principle a ranking score can depend on all the documents of the query; here for ease of presentation we consider the simple case in which it only depends on the corresponding document.)

Specifically, let $\{g_k(y_j, \mathbf{x})\}_{k=1}^{K_1}$ be a set of real-valued feature functions defined on document set $\mathbf{x}$ and label y_j ($j = 1, \ldots, m$), and $\{g'_k(y_u, y_v, \mathbf{x})\}_{k=1}^{K_2}$ be a set of real-valued feature functions defined on y_u, y_v, and $\mathbf{x}$ ($u, v = 1, \ldots, m, u \neq v$). Then C-CRF is a conditional probability distribution with the following density function,

$$\Pr(\mathbf{y}|\mathbf{x}) = \frac{1}{Z(\mathbf{x})} \exp\left\{\sum_{j}\sum_{k=1}^{K_1} \alpha_k g_k(y_j, \mathbf{x}) + \sum_{u,v}\sum_{k=1}^{K_2} \beta_k g'_k(y_u, y_v, \mathbf{x})\right\}, \qquad (6.7)$$

where α is a K_1-dimensional parameter vector and β is a K_2-dimensional parameter vector, and $Z(\mathbf{x})$ is a normalization function,

$$Z(\mathbf{x}) = \int_{\mathbf{y}} \exp\left\{\sum_{j}\sum_{k=1}^{K_1} \alpha_k g_k(y_j, \mathbf{x}) + \sum_{u,v}\sum_{k=1}^{K_2} \beta_k g'_k(y_u, y_v, \mathbf{x})\right\} d\mathbf{y}. \qquad (6.8)$$

Given training data $\{\mathbf{x}^{(i)}, \mathbf{y}^{(i)}\}_{i=1}^{n}$, the Maximum Likelihood Estimation can be used to estimate the parameters $\{\alpha, \beta\}$ of C-CRF. Specifically, the conditional log likelihood of the training data with respect to the C-CRF model can be computed as follows:

$$L(\alpha, \beta) = \sum_{i=1}^{n} \log \Pr(\mathbf{y}^{(i)}|\mathbf{x}^{(i)}; \alpha, \beta). \qquad (6.9)$$

Then the gradient ascent method is used to maximize the log likelihood, and the learned parameters are then used to rank the documents of a new query. Specifically, given documents $\mathbf{x}$ for the new query,

$$h(\mathbf{x}) = \arg\max_{\mathbf{y}} \Pr(\mathbf{y}|\mathbf{x}). \tag{6.10}$$

With the above general idea of C-CRF, in [7], it is demonstrated how to define specific forms of feature functions and the conditional probabilities for several relational ranking tasks, such as topic distillation and pseudo relevance feedback. In these specific cases, it has been shown in [7] that the optimal ranking functions have closed-form solutions. This greatly eases both training and test processes.

For example, the optimal ranking function for pseudo relevance feedback takes the following form:

$$F(\mathbf{x}) = \left(\alpha^T eI + \beta D - \beta R\right)^{-1} \mathbf{x}\alpha, \tag{6.11}$$

where e is a K_1-dimensional all-ones vector, I is an $m \times m$ identity matrix, R is a similarity matrix, and D is an $m \times m$ diagonal matrix with $D_{i,i} = \sum_j R_{i,j}$.

For another example, the optimal ranking function for topic distillation takes the following form:

$$h(\mathbf{x}) = \frac{1}{\alpha^T e} \left(2\mathbf{x}\alpha + \beta(D_r - D_c)e\right), \tag{6.12}$$

where D_r and D_c are two diagonal matrices with $D_{ri,i} = \sum_j R_{i,j}$ and $D_{ci,i} = \sum_j R_{j,i}$.

Experimental results on the LETOR benchmark datasets show that C-CRF can significantly outperform conventional learning-to-rank methods on the tasks of topic distillation and pseudo relevance feedback.

6.2 Learning Diverse Ranking

Ideally, the relational ranking framework introduced in the previous subsection can handle various relational ranking task by using the appropriate relationship to define its objective function. However, no concrete examples have been given in [6] and [7] on how the framework can be used to learn diverse ranking. On the other hand, there are several independent works that have investigated the problem of search result diversification. For example, in [9], Yue et al. investigate how to use structured SVM to predict diverse subsets. In [8], two online learning algorithms, the ranked explore and commit algorithm and the ranked bandits algorithm, are proposed to directly learn diverse ranking of documents based on users' click behaviors. In [1], a formal probabilistic formulation about search result diversification is proposed, in the context that the topical categories of both the query and the documents are given. In [4], axioms that any diversification system ought to satisfy are discussed. We will introduce [1] and [4] in more detail in this subsection.

In [1], it is investigated how to diversify search results by making explicit use of knowledge about the topics that the query or the documents may refer to. Specifically it is assumed that there exists a taxonomy of information, and the diversity in user intents is modeled at the topical level of the taxonomy. On this basis, an objective is proposed to explicitly trade off relevance and diversity as follows:

$$P(S|q) = \sum_c P(c|q) \left(1 - \prod_{x \in S} (1 - V(x|q, c)) \right), \qquad (6.13)$$

where $V(x|q, c)$ denotes the relevance probability of a document x for query q when the intended category is c.

It is not difficult to get the probabilistic interpretation of the above objective. Actually, it exactly describes the probability that at least one of the documents in the document set S satisfies the average user (in terms of expectation) who issues query q.

It has been proven that the general problem of optimizing the above objective is NP-hard. The good news is that the objective function admits a submodularity structure [12] that can be exploited for the implementation of a good approximation algorithm. Intuitively, a submodular function satisfies the economic principle of diminishing marginal returns, i.e., the marginal benefit of adding a document to a larger collection is less than that of adding it to a smaller collection. With this property, a greedy algorithm can be designed.

Given the top-k documents selected by some classical ranking algorithm for the target query, the greedy algorithm reorders these documents to maximize the objective $P(S|q)$. Let $U(c|q, S)$ denote the conditional probability that query q belongs to category c, given that all documents in set S fail to satisfy the user. Initially, before any document is selected, $U(c|q, \emptyset) = P(c|q)$. The algorithm selects output documents one at a time. At every step, it chooses the document that has the highest marginal utility defined as below,

$$g(x|q, c, S) = U(c|q, S)V(x|q, c). \qquad (6.14)$$

This marginal utility can be interpreted as the probability that the selected document satisfies the user given that all documents coming before it fail to do so. At the end of the loop, the conditional distribution is updated to reflect the inclusion of the new document to the result set using the Bayes rule.

As for the algorithm, the condition has been derived on which the greedy heuristic can lead to the optimal diversification result. Some further analysis shows that even if the condition is not met, the approximation error of the greedy method can be well bounded.

Inspired by the aforementioned new algorithm, the authors of [1] further propose refining existing evaluation measures in information retrieval to be intent-aware measures. For example, now NDCG and MAP will become

$$\text{NDCG}_{\text{IA}}@k = \sum_c P(c|q)\text{NDCG}@k, \qquad (6.15)$$

$$\text{MAP}_{\text{IA}}@k = \sum_{c} P(c|q)\text{MAP}@k. \tag{6.16}$$

Experimental results showed that the proposed method can significantly outperform previous methods in terms of the intent-aware measures.

In [4], it is formally pointed out that in most cases the diversification problem can be characterized as a bi-criteria optimization problem. That is, diversification can be viewed as combining both ranking (presenting more relevant results in the higher positions) and clustering (grouping documents satisfying similar intents) and therefore addresses a loosely defined goal of picking a set of the most relevant but novel documents. A very general objective function considering the above aspects can be formulated as follows.

$$L\big(S_k, q, f(\cdot), d(\cdot, \cdot)\big), \tag{6.17}$$

where S_k is the subset of documents presented at the top-k positions, q is the given query, $f(\cdot)$ is a relevance function, and $d(\cdot, \cdot)$ is a distance function.

Then a set of simple properties (called axioms) that any diversification system ought to satisfy are proposed as follows.[3]

- *Scale invariance:* Informally, this property states that the set selection function should be insensitive to the scaling of the input functions $f(\cdot)$ and $d(\cdot, \cdot)$.
- *Consistency:* Consistency states that making the output documents more relevant and more diverse, and making other documents less relevant and less diverse should not change the output of the ranking.
- *Richness:* Informally speaking, the richness condition states that we should be able to achieve any possible set as the output, given the right choice of relevance and distance functions.
- *Stability:* The stability condition seeks to ensure that the output set does not change arbitrarily with the output size. That is, the best top-$(k + 1)$ subset should be a super set of the best top-k subset.
- *Independence of irrelevant attributes:* This axiom states that the score of a set is not affected by most attributes of documents outside the set.
- *Monotonicity:* Monotonicity simply states that the addition of any document does not decrease the score of the set.
- *Strength of relevance:* This property ensures that no function L ignores the relevance function.
- *Strength of similarity:* This property ensures that no function L ignores the distance function.

It is pointed out in [4], however, that no function L can satisfy all the eight axioms. For example, the objective function in [1] violates the axioms of stability and

---

[3]Please note that these axioms might not be as necessary as claimed in some cases. For example, it is sometimes reasonable that a good diversification system does not possess the stability property, since the stability property somehow implies greedy systems which might not be optimal.

independence of irrelevant attributes. Therefore, one can only hope for diversification functions that satisfy a subset of the axioms. A few such examples are given as follows:

1. Max-sum diversification, which satisfies all the axioms, except stability.

$$L(S) = (k - 1) \sum_{u \in S} f(u) + 2\lambda \sum_{u,v \in S} d(u, v). \tag{6.18}$$

2. Max–min diversification, which satisfies all the axioms except consistency and stability.

$$L(S) = \min_{u \in S} f(u) + \lambda \min_{u,v \in S} d(u, v). \tag{6.19}$$

3. Mono-objective formulation, which satisfies all the axioms except consistency.

$$L(S) = \sum_{u \in S} f(u) + \frac{\lambda}{|U| - 1} \sum_{v \in U} d(u, v). \tag{6.20}$$

In addition to the two pieces of work introduced above, there are also many other works on learning diverse ranking. Actually, the task of learning diverse ranking has become a hot research topic in the research community. In 2009, the TREC conference even designed a special task for search result diversification. The goal of the diversity task is to return a ranked list of pages that together provide complete coverage for a query, while avoiding excessive redundancy in the result list.

In the task, 50 queries are used. Subtopics for each query are based on information extracted from the logs of a commercial search engine, and are roughly balanced in terms of popularity. Each topic is structured as a representative set of subtopics, each related to a different user need. Documents are judged with respect to the subtopics. For each subtopic, the human assessors will make a binary judgment as to whether or not the document satisfies the information need associated with the subtopic. α-NDCG [2] and MAP$_{IA}$ [1] are used as the evaluation measures. For more information, one can refer to the website of the task: http://plg.uwaterloo.ca/~trecweb/.

6.3 Discussions

In this chapter, we have introduced some existing works on relational ranking. While these works have opened a window to this novel task beyond conventional learning to rank, there are still many issues that need to be further investigated.

- As mentioned before, it is not clear how the general relational ranking framework can be used to solve the problem of learning diverse ranking. It would be interesting to look into this issue, so as to make the general relational ranking framework really general.

- In the algorithms introduced in this chapter, the diversity is modeled by the distance or dissimilarity between any two documents. However, this might be inconsistent with the real behavior of users. The reason is that users usually will not check every document in the result list, instead, they browse the ranking results in a top down manner. As a consequence, it would make more sense to define diversity as the relationship between a given document and all the documents ranked before it.

- In all the algorithms introduced in this chapter, the relationship is pairwise: either similarity, dissimilarity, or preference. Accordingly, a matrix (or graph) is used to model the relationship. However, in real applications, there are some other relationships that may go beyond "pairwise". For example, all the webpages in the same website have a multi-way relationship. It is more appropriate to use tensor (or hypergraph) to model such multi-way relationships. Accordingly, the relational ranking framework should be upgraded (note that tensor and hypergraph can contain matrix and graph as special cases).

References

1. Agrawal, R., Gollapudi, S., Halverson, A., Ieong, S.: Diversifying search results. In: Proceedings of the 2nd ACM International Conference on Web Search and Data Mining (WSDM 2009), pp. 5–14 (2009)
2. Clarke, C.L., Kolla, M., Cormack, G.V., Vechtomova, O., Ashkan, A., Buttcher, S., MacKinnon, I.: Novelty and diversity in information retrieval evaluation. In: Proceedings of the 31st Annual International ACM SIGIR Conference on Research and Development in Information Retrieval (SIGIR 2008), pp. 659–666 (2008)
3. Craswell, N., Hawking, D., Wilkinson, R., Wu, M.: Overview of the trec 2003 web track. In: Proceedings of the 12th Text Retrieval Conference (TREC 2003), pp. 78–92 (2003)
4. Gollapudi, S., Sharma, A.: An axiomatic approach for result diversification. In: Proceedings of the 18th International Conference on World Wide Web (WWW 2009), pp. 381–390 (2009)
5. Qin, T., Liu, T.Y., Zhang, X.D., Chen, Z., Ma, W.Y.: A study of relevance propagation for web search. In: Proceedings of the 28th Annual International ACM SIGIR Conference on Research and Development in Information Retrieval (SIGIR 2005), pp. 408–415 (2005)
6. Qin, T., Liu, T.Y., Zhang, X.D., Wang, D., Li, H.: Learning to rank relational objects and its application to web search. In: Proceedings of the 17th International Conference on World Wide Web (WWW 2008), pp. 407–416 (2008)
7. Qin, T., Liu, T.Y., Zhang, X.D., Wang, D.S., Li, H.: Global ranking using continuous conditional random fields. In: Advances in Neural Information Processing Systems 21 (NIPS 2008), pp. 1281–1288 (2009)
8. Radlinski, F., Kleinberg, R., Joachims, T.: Learning diverse rankings with multi-armed bandits. In: Proceedings of the 25th International Conference on Machine Learning (ICML 2008), pp. 784–791 (2008)
9. Yue, Y., Joachims, T.: Predicting diverse subsets using structural SVM. In: Proceedings of the 25th International Conference on Machine Learning (ICML 2008), pp. 1224–1231 (2008)

Chapter 7
Query-Dependent Ranking

Abstract In this chapter, we introduce query-dependent ranking. Considering the large differences between different queries, it might not be the best choice to use a single ranking function to deal with all kinds of queries. Instead, one may achieve performance gain by leveraging the query differences. To consider the query difference in training, one can use a query-dependent loss function. To further consider the query difference in the test process, a query-dependent ranking function is needed. Several ways of learning a query-dependent ranking function are reviewed in this chapter, including query classification-based approach, query clustering-based approach, nearest neighbor-based approach, and two-layer learning-based approach. Discussions are also made on the future research directions along this line.

Queries in web search may vary largely in semantics and the users' intentions they represent, in forms they appear, and in number of relevant documents they have in the document repository. For example, queries can be navigational, informational, or transactional. Queries can be personal names, product names, or terminology. Queries can be phrases, combinations of phrases, or natural language sentences. Queries can be short or long. Queries can be popular (which have many relevant documents) or rare (which only have a few relevant documents). If one can successfully leverage query differences in the process of learning to rank, one may have the opportunities to achieve better ranking performances in both training and test processes.

7.1 Query-Dependent Loss Function

In [12], the query differences are considered in the training process. This is achieved by using a query-dependent loss function. For ease of illustration, we assume that a regression-based algorithm is used for learning to rank, whose original loss function is defined as below.

$$L(f) = \sum_{i=1}^{n} \sum_{j=1}^{m^{(i)}} \left(y_j^{(i)} - f(x_j^{(i)}) \right)^2 + \lambda_f \| f \|^2. \tag{7.1}$$

T.-Y. Liu, *Learning to Rank for Information Retrieval*,
DOI 10.1007/978-3-642-14267-3_7, © Springer-Verlag Berlin Heidelberg 2011

The idea is to replace the loss term $(y_j^{(i)} - f(x_j^{(i)}))^2$ with the difference between the ground truth label $y_j^{(i)}$ and a query-dependent linear transformation of the ranking function, i.e., $g^{(i)}(f(x)) = \alpha^{(i)} f(x_j^{(i)}) + \beta^{(i)}$, where $\alpha^{(i)}$ and $\beta^{(i)}$ are parameters for query q_i. Accordingly, the loss function becomes

$$L(f; \alpha, \beta) = \sum_{i=1}^{n} \sum_{j=1}^{m^{(i)}} (y_j^{(i)} - \beta^{(i)} - \alpha^{(i)} f(x_j^{(i)}))^2$$

$$+ \lambda_\beta \|\beta\|^2 + \lambda_\alpha \|\alpha\|^2 + \lambda_f \|f\|^2, \qquad (7.2)$$

where $\beta = [\beta^{(i)}]_{i=1}^{n}$, $\alpha = [\alpha^{(i)}]_{i=1}^{n}$, and $\lambda_\alpha, \lambda_\beta, \lambda_f$ are regularization parameters.

By minimizing this loss function on the training data, one will get the optimal parameters $\{f, \alpha, \beta\}$. Note that in the test phase, only f is used. There are two reasons for doing so: (i) for new queries it is impossible to obtain parameter α and β; (ii) as long as the (unknown) parameter α is positive, the corresponding linear transformation on the ranking function will not change the ranking order of the documents.

In [2], a framework is proposed to better describe how to use query-dependent loss function. In particular, the loss function in training is formulated in the following general form:

$$L(f) = \sum_q L(f; q), \qquad (7.3)$$

where $L(f; q)$ is the query-level loss function defined on both query q and ranking function f, and each query has its own form of loss function.

Since it is infeasible to really give each individual query a different loss function, for practical consideration, the following query category-based realization of the above query-dependent loss function is proposed:

$$L(f) = \sum_q L(f; q) = \sum_q \left(\sum_c P(c|q) L(f; q, c) \right), \qquad (7.4)$$

where $L(f; q, c)$ denotes a category-level loss function defined on query q, ranking function f and q's category c.

The categorization of queries can be given according to the search intentions behind queries. For example, queries can be classified into three major categories according to Broder's taxonomy [3]: navigational, informational, and transactional. Under this taxonomy, a navigational query is intended to locate a specific Web page, which is often the official homepage or subpage of a site; an informational query seeks information on the query topic; and a transactional query seeks to complete a transaction on the Web. According to the definition of the above categories, for the navigational and transactional queries, the ranking model should aim to retrieve the exact relevant Web page for the top position in the ranking result; while for the informational query, the ranking model should target to rank more relevant Web pages on a set of top positions in the ranking result.

To incorporate this kind of query difference, a position-sensitive query-dependent loss function is proposed in [2]. In particular, for the navigational and transactional query, the loss function focuses on that exact relevant page; while for the informational query, the loss considers those relevant pages which should be ranked in the range of top-k positions.

Accordingly, the query-dependent loss can be written as follows.

$$L(f; q) = \alpha(q)L(f; q, c_I) + \big(1 - \alpha(q)\big)L(f; q, c_N), \tag{7.5}$$

where c_I denotes informational queries and c_N denotes navigational and transactional queries, $\alpha(q) = P(c_I|q)$ represents the probability that q is an informational query; $L(f; q, c)$ is defined with a position function $\Phi(q, c)$, i.e.,

$$L(f; q, c) = \sum_{j=1}^{m} L(f; x_j, y_j) I_{\{\pi_y(x_j) \in \Phi(q,c), \exists \pi_y \in \Omega_y\}}, \tag{7.6}$$

where Ω_y is the equivalent permutation set of the ground truth label, $\Phi(q, c)$ is a set of ranking positions on which users expect high result accuracy. Specifically, $\Phi(q, c_I) = \{1, \ldots, k\}$ and $\Phi(q, c_N) = \{1\}$.

In order to make the above query-dependent loss functions optimizable, a set of technologies are proposed, including learning the query classifier (based on a set of query features) and the ranking model (based on a set of query-document matching features) in a unified manner. The experimental results have shown that by applying the idea of query-dependent loss to existing algorithms like RankNet [4] and ListMLE [11], the ranking performances can be significantly improved.

7.2 Query-Dependent Ranking Function

In addition to the use of a query-dependent loss function, researchers have also investigated the use of different ranking functions for different (kinds of) queries.

7.2.1 Query Classification-Based Approach

In [9], Kang and Kim classify queries into categories based on search intentions and build different ranking models accordingly.

In particular, user queries are classified into two categories, the topic relevance task and the homepage finding task. In order to perform effective query classification, the following information is utilized:

- *Distribution difference:* Two datasets are constructed, one for topic relevance and the other for homepage finding. Then two language models are built from them. For a query, the difference in the probabilities given by the two models is used as a feature for query classification.

- *Mutual information difference:* Given a dataset, one can compute the mutual information (MI) to represent the dependency of terms in a query. Then the difference between the MI computed from the aforementioned two datasets is used as a feature.
- *Usage rate:* If query terms appear in title and anchor text frequently, the corresponding query does very possibly belong to the homepage finding task.
- *POS information:* Some topic relevance task queries include a verb to explain what the user wants to know. Therefore, by analyzing the part of speech information, one can extract features for query classification.

With the above features, one can use their linear combination to perform query classification.

After query classification, different algorithms and information are used for the queries belonging to different categories. For the topic relevance task, the content information (e.g., *TF* and *IDF*) is emphasized; on the other hand, for the homepage finding task, the link and URL information (e.g., PageRank, URL types (root, subroot, path, or file)) is emphasized.

Experimental results show that in this way, the ranking performance can be improved as compared to using a single ranking function.

7.2.2 K Nearest Neighbor-Based Approach

In [5], Geng et al. perform query-dependent ranking using the K Nearest Neighbor (KNN) method.

In the method, the training queries are positioned into the query feature space in which each query is represented by a feature vector. In the test process, given a query, its k nearest training queries are retrieved and used to learn a ranking model. Then this ranking model is used to rank the documents associated with the test query. This method is named KNN Online in [5]. This idea has been tested with Ranking SVM [7, 8], and the experimental results show significant performance improvements as compared to using one single ranking function and using query classification-based query-dependent ranking. The explanations on this experimental finding are as follows: (i) in the method ranking for a test query is conducted by leveraging the useful information of the similar training queries and avoiding the negative effects from the dissimilar ones; (ii) 'soft' classification of queries is carried out and similar queries are selected dynamically.

Despite the high ranking performance, one may have noticed a serious problem with the aforementioned method. Its time complexity is too high to use in practice. For each single test query, a KNN search plus a model training would be needed, both of which are very costly. In order to tackle this efficiency problem, two alternative methods have been developed, namely KNN Offline-1 and KNN Offline-2, which move the training offline and significantly reduce the test complexity.

KNN Offline-1 works in the following way (see Fig. 7.1). First, for each training query q_i, its k nearest neighbors, denoted as $N_k(q_i)$, in the query feature space are

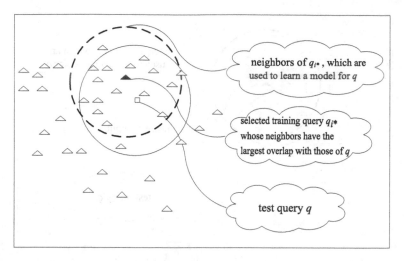

neighbors of q_{i^*}, which are used to learn a model for q

selected training query q_{i^*} whose neighbors have the largest overlap with those of q

test query q

Fig. 7.1 Illustration of KNN Offline-1

retrieved. Then, a model h_{q_i} is trained using the labeled data in $N_k(q_i)$ offline and in advance. In the test process, for a new query q, one first finds its k nearest neighbors $N_k(q)$, and then compares $N_k(q)$ with every $N_k(q_i)$ ($i = 1, \ldots, n$) so as to find the one sharing the largest number of instances with $N_k(q)$. Suppose it is $N_k(q_{i^*})$. Then $h_{q_{i^*}}$, the model learned from the selected neighborhood $N_k(q_{i^*})$ (it has been created offline and in advance), is used to rank the documents of query q.

KNN Offline-2 works as follows (see Fig. 7.2). First, for each training query q_i, its k nearest neighbors $N_k(q_i)$ in the query feature space are retrieved. Then, a model h_{q_i} is trained using the labeled data in $N_k(q_i)$ offline and in advance. Instead of seeking the k nearest neighbors for the test query q, one only finds its single nearest neighbor in the query feature space. Supposing this nearest neighbor to be q_{i^*}, then the model $h_{q_{i^*}}$ trained from $N_k(q_{i^*})$ is used to test query q. In this way, one further simplifies the search of k nearest neighbors to that of a single nearest neighbor, and no longer needs to compute the similarity between neighborhoods.

It has been shown that the test complexities of KNN Offline-1 and KNN Offline-2 are both $O(n \log n)$, which is comparable with the single model approach. At the same time, experimental results in [5] show that the effectiveness of these two offline algorithms are comparable with KNN Online. Actually, it has been proven that the approximations of KNN Online by KNN Offline-1 and KNN Offline-2 are accurate in terms of difference in loss of prediction, as long as the learning algorithm used is stable with respect to minor changes in training examples. And learning-to-rank algorithms with regularization items in their loss functions, such as Ranking SVM, are typical stable algorithms.

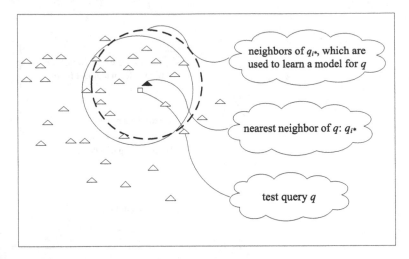

Fig. 7.2 Illustration of KNN Offline-2

7.2.3 Query Clustering-Based Approach

In [1], a similar idea to that of [5] is used to perform query-dependent ranking. The difference lies in that in [1] query clustering is used instead of the KNN techniques. Specifically, training queries are clustered offline. For this purpose, query q is represented by the point cloud of its associated documents in the feature space, denoted as D_q. Then for two queries q and q', one first conducts PCA to compute the sequences of the principle components of D_q and $D_{q'}$, denoted as $(u_1, \ldots, u_P)$ and $(u'_1, \ldots, u'_P)$ respectively. Then the similarity between two queries is computed as the sum of the inner product between two corresponding principle components:

$$\text{sim}(q, q') = \frac{1}{P} \sum_{p=1}^{P} |u_p^T u'_p|. \tag{7.7}$$

With this query similarity, all the training queries are clustered using complete-link agglomerative clustering [10] into C clusters $\{Q_1, \ldots, Q_C\}$. For each cluster Q_k, a model is trained using queries belonging to it. During test time, given a test query q, one locates the nearest training cluster to the test query as follows:

$$\arg \max_{k} \max_{q' \in Q_k} \text{sim}(q, q'). \tag{7.8}$$

After that one uses the model trained from that cluster to rank the documents associated with the test query.

7.2.4 Two-Layer Learning Approach

In [6], He and Liu argued that no matter whether query categorization, clustering, or nearest neighbors is being used, one basically only pre-trains a finite number of models and selects the most suitable one from them to rank the documents for a test query. However, given the extremely large space of queries, this "finite-model" solution might not work well for all the test queries. To solve the problem, they propose a novel solution with an "infinite granularity".

Specifically, the training process of query-dependent ranking is defined as the following optimization problem:

$$\min_{v} \sum_{i=1}^{n} L\big(\hat{\mathbf{y}}^{(i)}, \mathbf{y}^{(i)}\big),$$

$$\text{s.t.} \quad \hat{\mathbf{y}}^{(i)} = \text{sort} \circ f\big(w^{(i)}, \mathbf{x}^{(i)}\big),$$

$$w^{(i)} = g\big(v, z^{(i)}\big),$$

where $f(\cdot)$ is the scoring function whose parameter is $w = g(v, z)$, in which z is the query feature vector, and v is the parameter of function g.

From the above formulation, it can be seen that there are two layers of learning in the training process:

- The first layer is named as the *document layer*. This layer corresponds to learning the parameter w in $f(w, \mathbf{x})$. In other words, one learns how to generate the ranking scores for the documents by combining their document features. The document-layer learning is the same as the learning process in conventional learning to rank.
- The second layer is named as the *query layer*. This layer regards the parameter w as the target, and tries to generate w by combining query features, i.e., $w = g(v, z)$. In other words, one no longer regards w as a deterministic vector, but a function of the query features. Then, for different queries, since their query features are different, their corresponding ranking models will therefore become different.

The test process is also different from conventional learning to rank. Given a test query q, there is no pre-determined ranking model w that can be used to rank its associated documents. Instead one needs to generate such a model on the fly. That is, one first obtains the query features z, and then uses the learned parameter v to create a ranking model $w = g(v, z)$. This model is then used to rank the documents associated with q based on their document features.

As compared to other query-dependent ranking methods, this two-layer method has the following properties.

- In this approach, the way of using the query features are learned according to the relevance label and loss functions for ranking. In contrast, the way of using the query features in previous methods is either unsupervised or not according to

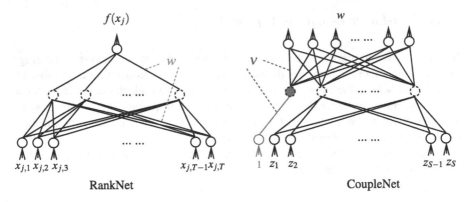

Fig. 7.3 Two-layer approach to query-dependent ranking

relevance labels. In this regard, this approach can avoid the mismatch between query clustering/classification with the goal of ranking.

- In this approach, query features are directly used to adjust the ranking model, instead of through a finite number of query clusters or categories. In this regard, this approach is more direct, and can achieve query-dependent ranking at a much finer granularity than previous work.

Note that this two-layer learning approach is very general and can be applied to many start-of-the-art learning-to-rank algorithms. In [6], He and Liu take RankNet as an example to demonstrate the use of the approach. Specifically, the original RankNet algorithm is used for document-layer learning, and a "couple" neural network is created to perform the query-layer learning (denoted as CoupleNet). The corresponding two-layer learning process can be illustrated using Fig. 7.3. Experimental results reported in [6] show that this two-layer learning approach can significantly outperform both conventional learning-to-rank methods and previous query-dependent ranking methods.

7.3 Discussions

In this chapter, we have introduced query-dependent ranking methods. Although they have shown promising experimental results, we would like to point out that there are still several important issues to consider along this research direction.

- Most of the methods rely on query features. The query features are used to perform query clustering, classification, and to find nearest neighbors. In the ideal case, the queries in the same cluster, category, or close to each other, should share similar ranking models. However, the reality is that the query features might not be able to reflect the similarity between ranking models. In other words, there is a kind of "gap" between the information we have (query features) and our goal (model-based clustering or classification). Note that the two-layer approach can avoid this problem to a certain degree.

- In most of the methods introduced in this chapter, the number of parameters (to learn) is much larger than that for the conventional ranking methods. As a result, the learning process in these methods might be easy to over fit. It is critical to control the risk of over-fitting.
- In the above descriptions, we have mainly focused on the algorithmic challenges for query-dependent ranking. In addition to this, there are also many engineering challenges. For example, in many methods, one needs to extract query features on the fly, and then generate a ranking model or select the most appropriate ranking model from a large number of candidates based on query features. This obviously increases the response time of search engines. If one cannot find a highly efficient way to solve this problem, the user experience will be hurt.

References

1. Banerjee, S., Dubey, A., Machichihar, J., Chakrabarti, S.: Efficient and accurate local learning for ranking. In: SIGIR 2009 Workshop on Learning to Rank for Information Retrieval (LR4IR 2009) (2009)
2. Bian, J., Liu, T.Y., Qin, T., Zha, H.: Ranking with query-dependent loss for web search. In: Proceedings of the 3rd International Conference on Web Search and Web Data Mining (WSDM 2010), pp. 141–150 (2010)
3. Broder, A.: A taxonomy of web search. SIGIR Forum **36**(2), 3–10 (2002)
4. Burges, C.J., Shaked, T., Renshaw, E., Lazier, A., Deeds, M., Hamilton, N., Hullender, G.: Learning to rank using gradient descent. In: Proceedings of the 22nd International Conference on Machine Learning (ICML 2005), pp. 89–96 (2005)
5. Geng, X.B., Liu, T.Y., Qin, T., Li, H., Shum, H.Y.: Query-dependent ranking using k-nearest neighbor. In: Proceedings of the 31st Annual International ACM SIGIR Conference on Research and Development in Information Retrieval (SIGIR 2008), pp. 115–122 (2008)
6. He, Y., Liu, T.Y.: Query-dependent ranking using couplenet. Tech. rep., Microsoft Research (2010)
7. Herbrich, R., Obermayer, K., Graepel, T.: Large margin rank boundaries for ordinal regression. In: Advances in Large Margin Classifiers, pp. 115–132 (2000)
8. Joachims, T.: Optimizing search engines using clickthrough data. In: Proceedings of the 8th ACM SIGKDD International Conference on Knowledge Discovery and Data Mining (KDD 2002), pp. 133–142 (2002)
9. Kang, I., Kim, G.: Query type classification for web document retrieval. In: Proceedings of the 26th Annual International ACM SIGIR Conference on Research and Development in Information Retrieval (SIGIR 2003), pp. 64–71 (2003)
10. Manning, C.D., Raghavan, P., Schutze, H.: Introduction to Information Retrieval. Cambridge University Press, Cambridge (2008)
11. Xia, F., Liu, T.Y., Wang, J., Zhang, W., Li, H.: Listwise approach to learning to rank—theorem and algorithm. In: Proceedings of the 25th International Conference on Machine Learning (ICML 2008), pp. 1192–1199 (2008)
12. Zha, H., Zheng, Z., Fu, H., Sun, G.: Incorporating query difference for learning retrieval functions in world wide web search. In: Proceedings of the 15th ACM International Conference on Information and Knowledge Management (CIKM 2006), pp. 307–316 (2006)

Chapter 8
Semi-supervised Ranking

Abstract In this chapter, we introduce semi-supervised learning for ranking. The motivation of this topic comes from the fact that we can always collect a large number of unlabeled documents or queries at a low cost. It would be very helpful if one can leverage such unlabeled data in the learning-to-rank process. In this chapter, we mainly review a transductive approach and an inductive approach to this task, and discuss how to improve these approaches by taking the unique properties of ranking into consideration.

So far in the previous chapters of the book, we have mainly discussed supervised learning in ranking. However, just like the case in classification, sometimes unlabeled data will help us reduce the volume of required labeled data. There have been some preliminary attempts [1, 2] on semi-supervised ranking.

8.1 Inductive Approach

In [1], an inductive approach is taken. More specifically, the ground-truth labels of the labeled documents are propagated to the unlabeled documents, according to their mutual similarity in the feature space. The same technology has been widely used in semi-supervised classification [4, 7, 8].

In order to exploit information from the unlabeled dataset, it is assumed that an unlabeled document that is similar to a labeled document should have similar label to that labeled document. One begins with selecting unlabeled documents that are the most similar to a labeled document x and assign them the corresponding relevance judgment y. For ease of discussion, we refer to such unlabeled documents as *automatically-labeled documents*, while the original labeled documents *human-labeled documents*.

After the label propagation, a simple approach is to add these automatically-labeled documents to the original training set and then learn a ranking function as in the supervised case, e.g., using RankBoost [3]. However, this training scheme suffers from the following drawback. As the automatically-labeled documents have error-prone labels, the ranking performance would be highly dependent on how ro-

T.-Y. Liu, *Learning to Rank for Information Retrieval*,
DOI 10.1007/978-3-642-14267-3_8, © Springer-Verlag Berlin Heidelberg 2011

bust the training scheme is to noisy labels. Therefore, instead of mixing the human-labeled and automatically-labeled documents, it is better to treat the two types of documents separately. That is, when using RankBoost, document pairs are constructed only within the same type of documents, and separate distributions are maintained for the two types of document pairs. In this way, the two types of data will contribute separately to the overall loss function (a parameter λ is used to trade off the losses corresponding to different types of documents). It has been proven that with such a treatment the training process can still converge, and some original nice properties of RankBoost can be inherited.

The proposed method has been tested on a couple of bipartite ranking tasks, with AUC (the area under the ROC curve) as the evaluation measure. The experimental results have shown that the proposed approach can improve the accuracy of bipartite ranking, and even when there is only a small amount of labeled data, the performance can be very good by well utilizing the unlabeled data.

8.2 Transductive Approach

In [2], a transductive approach is taken; the key idea is to automatically derive better features using the unlabeled test data to improve the effectiveness of model training. In particular, an unsupervised learning method (specifically, kernel PCA in [2]) is applied to discover salient patterns in each list of retrieved test documents. In total four different kernels are used: polynomial kernel, RBF kernel, diffusion kernel, and linear kernel (in this specific case, kernel PCA becomes exactly PCA). The training data are then projected onto the directions of these patterns and the resulting numerical values are added as new features. The main assumption in this approach is that this new training set (after projection) better characterizes the test data, and thus should outperform the original training set when learning rank functions. RankBoost [3] is then taken as an example algorithm to demonstrate the effectiveness of this transductive approach.

Extensive experiments on the LETOR benchmark datasets (see Chap. 10) have been conducted in [2] to test the effectiveness of the proposed transductive approach. The experimental results show that the ranking performance can be improved by using the unlabeled data in this way. At the same time, detailed analyses have been performed on some issues related to this semi-supervised learning process, e.g., whether the Kernel PCA features can be interpreted, whether non-linear Kernel PCA helps, how the performance varies across queries, and what the computational complexity is. The general conclusions are as follows.

- Kernel PCA features are in general difficult to interpret and most Kernel PCA features have little correlation to the original features.
- Non-linearity is important in most cases, but one should not expect non-linear kernels to always outperform linear ones. The best strategy is to employ multiple kernels.

- The proposed transductive approach does not give improvements across all queries. Gains come from the greater proportion of improved queries than that of degraded queries.
- The transductive approach requires online computations. The total time per query is around hundreds of seconds for the datasets under investigation. Therefore, it is desired that better code optimization or novel distributed algorithms can be used to make the approach practically applicable.

8.3   Discussions

As we can see, the above works have borrowed some concepts and algorithms from semi-supervised classification. Although good ranking performances have been observed, the validity of doing so may need further justification. For example, since similarity is essential to many classification algorithms (i.e., "similar documents should have the same class label"), it looks very natural and reasonable to propagate labels cross similar documents. However, in ranking, similarity does not play the same central role. It seems that preference is more fundamental than similarity. Then the question is whether it is still natural and reasonable to conduct similarity-based label propagation for semi-supervised ranking.

Furthermore, in classification, if we do not have class labels, we know nothing about the conditional probability $p(y|x)$. However, in ranking, even if we do not have ground-truth labels, we still have several very strong rankers, such as BM25 [6] and LMIR [5], which can give us a relatively reasonable guess on which document should be ranked higher. In other words, we have some knowledge about the unlabeled data. If we can incorporate such knowledge into the semi-supervised ranking process, we may have the chance to do a better job.

References

1. Amini, M.R., Truong, T.V., Goutte, C.: A boosting algorithm for learning bipartite ranking functions with partially labeled data. In: Proceedings of the 31st Annual International ACM SIGIR Conference on Research and Development in Information Retrieval (SIGIR 2008), pp. 99–106 (2008)
2. Duh, K., Kirchhoff, K.: Learning to rank with partially-labeled data. In: Proceedings of the 31st Annual International ACM SIGIR Conference on Research and Development in Information Retrieval (SIGIR 2008), pp. 251–258 (2008)
3. Freund, Y., Iyer, R., Schapire, R., Singer, Y.: An efficient boosting algorithm for combining preferences. Journal of Machine Learning Research 4, 933–969 (2003)
4. Niu, Z.Y., Ji, D.H., Tan, C.L.: Word sense disambiguation using label propagation based semi-supervised learning. In: Proceedings of the 403rd Annual Meeting of the Association for Computational Linguistics (ACL 2005), pp. 395–402 (2005)
5. Ponte, J.M., Croft, W.B.: A language modeling approach to information retrieval. In: Proceedings of the 21st Annual International ACM SIGIR Conference on Research and Development in Information Retrieval (SIGIR 1998), pp. 275–281 (1998)
6. Robertson, S.E.: Overview of the okapi projects. Journal of Documentation 53(1), 3–7 (1997)

7. Tong, W., Jin, R.: Semi-supervised learning by mixed label propagation. In: Proceedings of the 22nd AAAI Conference on Artificial Intelligence (AAAI 2007), pp. 651–656 (2007)
8. Xiaojin Zhu, Z.G.: Learning from labeled and unlabeled data with label propagation. Ph.D. thesis, Carnegie Mellon University (2002)

Chapter 9
Transfer Ranking

Abstract In this chapter, we introduce transfer learning for ranking, or transfer ranking for short. Transfer ranking is a task to transfer the knowledge contained in one learning-to-rank dataset or problem to another learning-to-rank dataset or problem. This is a typical task in real ranking applications, e.g., transferring from an old training set to a new one, or transferring from one market to another. In this chapter, we will briefly review the feature-level transfer ranking approach and the instance-level transfer ranking approach, and discuss the future research directions along this line.

Transfer learning is concerned with using a related but slightly different auxiliary task to help a given learning task.[1] The concept of transfer learning can be applied to many real ranking scenarios. Here we give some of them as examples.

- When training a ranking model for a search engine, large amounts of query-document pairs were previously labeled. After a period of time, these data may become outdated, since the distribution of queries submitted by users is time-varying. The prediction on these outdated data can serve as an auxiliary task to the prediction on the new data.
- Suppose a ranking model is desired for a newly born vertical search engine while only labeled data from another vertical search engine are available. Then the prediction on the old vertical search data is an auxiliary task to the prediction on the new vertical search data.
- Commercial search engines usually collect a large number of labeled data for their main-stream market, however, they cannot do the same thing for every niche market. To tackle this problem, many search engines leverage the training data from the main-stream market to make the ranking model for the niche market more effective and reliable. It is clear that the prediction on the main-stream market data is an auxiliary task to the prediction on the niche-market data.

Usually, in transfer learning, the (small amount of) labeled data for the given learning are is called the "target-domain data", while the labeled data associated

---

[1] See http://socrates.acadiau.ca/courses/comp/dsilver/NIPS95_LTL/transfer.workshop.1995.html.

T.-Y. Liu, *Learning to Rank for Information Retrieval*,
DOI 10.1007/978-3-642-14267-3_9, © Springer-Verlag Berlin Heidelberg 2011

with the auxiliary task are called the "source-domain data". Evidently, the source-domain data cannot be utilized directly in training a ranking model for the target domain due to different data distributions. And this is exactly what the technology of transfer learning wants to address.

Some preliminary attempts on transfer ranking have been made in recent years, such as [1] and [4]. Here, we take [1] as an example to illustrate the detailed process of transfer ranking. In particular, the authors of [1] have studied the problem of transfer ranking from two aspects, i.e., feature level and instance level, respectively.

9.1 Feature-Level Transfer Ranking

The *feature-level transfer ranking* method assumes that there exists a low-dimensional feature representation (may be transformations of the original features) shared by both source-domain and target-domain data. Specifically, suppose that these common features are the linear combinations of the original features,

$$z_k = u_k^T x, \tag{9.1}$$

where x is the original feature vector and u_k is the regression parameters for the new feature z_k. We use U to denote the matrix whose columns correspond to the vectors u_k.

Suppose the scoring function f is linear, that is, $f(z) = \alpha^T z$, where z is the learned common features. As a result, we have $f(z) = \alpha^T U x = w^T x$, where $w = \alpha U^T$. Let $A = [\alpha_s, \alpha_t]$, and $W = [w_s, w_t]$, where s and t stand for source domain and target domain, respectively, then we have $W = UA$.

The next step is to minimize the losses for both the source and target domains, in a manner of multi-objective optimization, in order to learn U and A simultaneously. Intuitively, according to the assumption that the source domain and target domain share a few common features, A should have some rows which are identically equal to zero. In order to reflect this intuition, a $(2, 1)$-regularization item for A is added, which is obtained by first computing the L_2 norm of the rows of matrix A to form a new vector, and then computing the L_1 norm of this vector. It has been shown in [1] that although this optimization problem itself is non-convex, it can be converted to an equivalent convex optimization problem and get effectively solved.

According to the experimental results in [1], the feature-level transfer ranking method can gain improvements over baselines, such as Ranking SVM [2, 3].

9.2 Instance-Level Transfer Ranking

The *instance-level transfer ranking* method makes use of the source-domain data by adapting each of them to the target domain from a probabilistic distribution point of view. Technically, this is implemented via instance re-weighting. Suppose the source and target domains share the same set of features, and their differences only lie in the different distributions of the data. Denote the source and target distributions by P_s and P_t, respectively, then we have

$$w_t = \arg\min \int L(w; \mathbf{x}, \mathbf{y}) P_t(\mathbf{x}, \mathbf{y}) \, d\mathbf{x} \, d\mathbf{y}$$

$$= \arg\min \int \frac{P_t(\mathbf{x}, \mathbf{y})}{P_s(\mathbf{x}, \mathbf{y})} L(w; \mathbf{x}, \mathbf{y}) P_s(\mathbf{x}, \mathbf{y}) \, d\mathbf{x} \, d\mathbf{y}$$

$$= \arg\min \int \frac{P_t(\mathbf{x})}{P_s(\mathbf{x})} \frac{P_t(\mathbf{y}|\mathbf{x})}{P_s(\mathbf{y}|\mathbf{x})} L(w; \mathbf{x}, \mathbf{y}) P_s(\mathbf{x}, \mathbf{y}) \, d\mathbf{x} \, d\mathbf{y}. \qquad (9.2)$$

Let $\delta = \frac{P_t(\mathbf{y}|\mathbf{x})}{P_s(\mathbf{y}|\mathbf{x})}$ and $\eta = \frac{P_t(\mathbf{x})}{P_s(\mathbf{x})}$; one obtains

$$w_t = \arg\min_w \int \delta \eta L(w; \mathbf{x}, \mathbf{y}) P_s(\mathbf{x}, \mathbf{y}) \, d\mathbf{x} \, d\mathbf{y}. \qquad (9.3)$$

In other words, with re-weighting factors, the minimization of the loss on the source domain can also lead to the optimal ranking function on the target domain. Therefore, in practice, w_t can be learned by minimizing the re-weighted empirical risk on the source-domain data:

$$w_t = \arg\min_w \sum_{i=1}^{n_s} \delta_i \eta_i L\left(w; \mathbf{x}_s^{(i)}, \mathbf{y}_s^{(i)}\right), \qquad (9.4)$$

where the subscript s means the source domain, and n_s is the number of queries in the source-domain data.

In [1], it is assumed that $\eta_i = 1$. In other words, there is no difference in the distribution of $\mathbf{x}$ for the source and target domains. To set δ_i, the following heuristic method is used. First a ranking model is trained from the target-domain data and then it is tested on the source-domain data. If a pair of documents in the source-domain data is ranked correctly, the corresponding pair is retained and assigned with a weight; else, it is discarded. Since in learning to rank each document pair is associated with a specific query, the pairwise precision of this query is used to determine δ_i:

$$\delta_i = \frac{\# \text{ pairs correctly ranked of a query}}{\# \text{ total pairs of a query}}.$$

According to the experimental results in [1], the instance-level method only works well for certain datasets. On some other datasets, its performance is even worse than only using the target-domain data. This in a sense shows that simple re-weighting might not effectively bridge the gap between the source and target domains.

9.3 Discussions

In this chapter, we have introduced some transfer ranking methods. As for the future work on transfer ranking, the following issues may need to be investigated.

- The effectiveness of transfer learning heavily replies on the relation between the source and target tasks. Researchers usually use the term "relatedness" to describe the relation. In the literature, there are some meaningful definitions for the relatedness of classification tasks, however, it is still unclear how the relatedness for ranking should be defined.
- As can be seen, the existing methods on transfer ranking, no matter feature level or instance level, are pretty much similar to the transfer learning methods for classification. It is not clear what kind of impact that the differences between ranking and classification would make on the problem. It is expected that some novel transfer learning methods, which are tailored more towards the unique properties of ranking, may lead to better results.
- In fact, these different levels of transfer learning are complementary, how to effectively combine them into one generic framework is an open question.
- Transfer learning is highly related to multi-task learning. Therefore, in addition to studying transfer ranking, it makes sense to further look at the topics like multi-task ranking. But again, it is a key problem whether one can effectively leverage the unique properties of ranking when moving forward in this direction.

References

1. Chen, D., Yan, J., Xiong, Y., Xue, G.R., Wang, G., Chen, Z.: Knowledge transfer for cross domain learning to rank. Information Retrieval Journal. Special Issue on Learning to Rank 13(3), doi10.1007/s10791-009-9111-2 (2010)
2. Herbrich, R., Obermayer, K., Graepel, T.: Large margin rank boundaries for ordinal regression. In: Advances in Large Margin Classifiers, pp. 115–132 (2000)
3. Joachims, T.: Optimizing search engines using clickthrough data. In: Proceedings of the 8th ACM SIGKDD International Conference on Knowledge Discovery and Data Mining (KDD 2002), pp. 133–142 (2002)
4. Wu, Q., Burges, C., Svore, K., Gao, J.: Ranking, boosting, and model adaptation. Information Retrieval Journal. Special Issue on Learning to Rank 13(3), doi10.1007/s10791-009-9112-1 (2010)

Part IV
Benchmark Datasets for Learning to Rank

In this part, we introduce the publicly available benchmark datasets for learning to rank, which provide good experimental environments for researchers to study different kinds of learning to rank algorithms. We first make a detailed introduction to the LETOR datasets, the first benchmark datasets for the research on learning to rank. Then we present the experimental results for typical learning to rank algorithms provided with the LETOR datasets and give some discussions on these algorithms. After that, we briefly mention two newly released benchmark datasets, which are much larger than LETOR and can support more types of learning to rank research.

After reading this part, the readers are expected to be familiar with existing benchmark datasets, and know how to test their own algorithms on these datasets.

Part IV
Benchmark Datasets for Learning to Rank

Chapter 10
The LETOR Datasets

Abstract In this chapter, we introduce the LETOR benchmark datasets, including the following aspects: document corpora (together with query sets), document sampling, feature extraction, meta information, cross validation, and major ranking tasks supported.

10.1 Overview

As we all know, a benchmark dataset with standard features and evaluation measures is very helpful for the research on machine learning. For example, there are benchmark datasets such as Reuters[1] and RCV-1[2] for text classification, and UCI[3] for general machine learning. However, there were no such benchmark datasets for ranking until the LETOR datasets [8] were released in early 2007. In recent years, the LETOR datasets have been widely used in the experiments of learning-to-rank papers, and have helped to greatly move forward the research on learning to rank. Up to the writing of this book, there have been several versions released for LETOR. In this chapter, we will mainly introduce two most popularly used versions, LETOR 3.0 and 4.0. In particular, we will describe the details of these datasets including document corpus, document sampling, feature extraction, meta information, and learning tasks.

10.2 Document Corpora

Three document corpora together with nine query sets are used in the LETOR datasets. The first two document corpora are used in LETOR 3.0, while the third one is used in LETOR 4.0.

---

[1] http://www.daviddlewis.com/resources/testcollections/reuters21578/.

[2] http://jmlr.csail.mit.edu/papers/volume5/lewis04a/lyrl2004_rcv1v2_README.htm.

[3] http://archive.ics.uci.edu/ml/.

T.-Y. Liu, *Learning to Rank for Information Retrieval*,
DOI 10.1007/978-3-642-14267-3_10, © Springer-Verlag Berlin Heidelberg 2011

Table 10.1 Number of
queries in TREC Web track

Task	TREC2003	TREC2004
Topic distillation	50	75
Homepage finding	150	75
Named page finding	150	75

10.2.1 The "Gov" Corpus and Six Query Sets

In TREC 2003 and 2004, a special track for web information retrieval, named the
Web track,[4] was organized. The track used the "Gov" corpus, which is based on a
January, 2002 crawl of the .gov domain. There are in total 1,053,110 html documents
in this corpus.

There are three search tasks in the Web track: topic distillation (TD), homepage
finding (HP), and named page finding (NP). *Topic distillation* aims to find a list of
entry points for good websites principally devoted to the topic. *Homepage finding*
aims at returning the homepage of the query. *Named page finding* aims to return the
page whose name is exactly identical to the query. Generally speaking, there is only
one answer for homepage finding and named page finding. The numbers of queries
in these three tasks are shown in Table 10.1. For ease of reference, we denote the
query sets in these two years as TD2003, TD2004, HP2003, HP2004, NP2003, and
NP2004, respectively.

Due to the large scale of the corpus, it is not feasible to check every document
and judge its relevance to a given query. The practice in TREC is as follows. Given
a query, only some "possibly" relevant documents, which are ranked high in the
runs submitted by the participants, are selected for labeling. Given a query, human
assessors are asked to label whether these possibly relevant documents are really
relevant. All the other documents, including those checked but not labeled as relevant by the human assessors and those not ranked high in the submitted runs at all,
are regarded as irrelevant in the evaluation process [2].

Many research papers [11, 13, 19, 20] have used the three tasks on the "Gov"
corpus as their experimental platform.

10.2.2 The OHSUMED Corpus

The OHSUMED corpus [5] is a subset of MEDLINE, a database on medical pub-
lications. It consists of 348,566 records (out of over 7,000,000) from 270 medical
journals during the period of 1987–1991. The fields of a record include title, abstract, MeSH indexing terms, author, source, and publication type.

---

[4]http://trec.nist.gov/tracks.html.

A query set with 106 queries on the OHSUMED corpus has been used in many previous works [11, 19], with each query describing a medical search need (associated with patient information and topic information). The relevance degrees of the documents with respect to the queries are judged by human assessors, on three levels: definitely relevant, partially relevant, and irrelevant. There are a total of 16,140 query–document pairs with relevance judgments.

10.2.3 The "Gov2" Corpus and Two Query Sets

The Million Query (MQ) track ran for the first time in TREC 2007 and then became a regular track in the following years. There are two design purposes of the MQ track. First, it is an exploration of ad-hoc retrieval on a large collection of documents. Second, it investigates questions of system evaluation, particularly whether it is better to evaluate using many shallow judgments or fewer thorough judgments.

The MQ track uses the so-called "terabyte" or "Gov2" corpus as its document collection. This corpus is a collection of Web data crawled from websites in the .gov domain in early 2004. This collection includes about 25,000,000 documents in 426 gigabytes.

There are about 1700 queries with labeled documents in the MQ track of 2007 (denoted as MQ2007 for short) and about 800 queries in the MQ track of 2008 (denoted as MQ2008). The judgments are given in three levels, i.e., highly relevant, relevant, and irrelevant.

10.3 Document Sampling

Due to a similar reason to selecting documents for labeling, it is not feasible to extract feature vectors of all the documents in a corpus either. A reasonable strategy is to sample some "possibly" relevant documents, and then extract feature vectors for the corresponding query–document pairs.

For TD2003, TD2004, NP2003, NP2004, HP2003, and HP2004, following the suggestions in [9] and [12], the documents are sampled in the following way. First, the BM25 model is used to rank all the documents with respect to each query, and then the top 1000 documents for each query are selected for feature extraction. Please note that this sampling strategy is to ease the experimental investigation, and this is by no means to say that learning to rank can only be applicable in such a re-ranking scenario.

Different from the above tasks in which unjudged documents are regarded as irrelevant, in OHSUMED, MQ2007, and MQ2008, the judgments explicitly contain the category of "irrelevant" and the unjudged documents are ignored in the evaluation. Correspondingly, in LETOR, only judged documents are used for feature extraction and all the unjudged documents are ignored for these corpora.

There are some recent discussions on the document sampling strategies for learning to rank, such as [1]. It is possible that different sampling strategies will lead to different effectiveness in training, however, currently these strategies have not been applied in the LETOR datasets.

10.4 Feature Extraction

In this section, we introduce the feature representation of documents in LETOR. The following principles are used in the feature extraction process.

1. To cover as many classical features in information retrieval as possible.
2. To reproduce as many features proposed in recent SIGIR papers as possible, which use the OHSUMED, "Gov", or "Gov2" corpus for their experiments.
3. To conform to the settings in the original papers.

For the "Gov" corpus, 64 features are extracted for each query–document pair, as shown in Table 10.2. Some of these features are dependent on both the query and the document, some only depend on the document, and some others only depend on the query. In the table, q represents a query, which contains terms $t_1, \ldots, t_M$; $TF(t_i, d)$ denotes the number of occurrences of query term t_i in document d. Note that if the feature is extracted from a stream (e.g., title, or URL), $TF(t_i, d)$ means the number of occurrences of t_i in the stream.

From the above table, we can find many classical information retrieval features, such as term frequency and BM25 [16]. At the same time, there are also many features extracted according to recent SIGIR papers. For example, Topical PageRank and Topical HITS are computed according to [10]; sitemap and hyperlink based score/feature propagations are computed according to [17] and [13], HostRank is computed according to [20], and extracted title is generated according to [6]. For more details about the features, please refer to the LETOR website http://research.microsoft.com/~LETOR/.

For the OHSUMED corpus, 45 features are extracted in total, as shown in Table 10.3. In the table, $|C|$ means the total number of documents in the corpus. For more details of these features, please refer to the LETOR website.

For the "Gov2" corpus, 46 features are extracted as shown in Table 10.4. Again, more details about these features can be found at the LETOR website.

10.5 Meta Information

In addition to the features, the following meta information has been provided in LETOR.

- Statistical information about the corpus, such as the total number of documents, the number of streams, and the number of (unique) terms in each stream.

Table 10.2 Features for the "Gov" corpus	ID	Feature description
	1	$\sum_{t_i \in q \cap d} TF(t_i, d)$ in body
	2	$\sum_{t_i \in q \cap d} TF(t_i, d)$ in anchor
	3	$\sum_{t_i \in q \cap d} TF(t_i, d)$ in title
	4	$\sum_{t_i \in q \cap d} TF(t_i, d)$ in URL
	5	$\sum_{t_i \in q \cap d} TF(t_i, d)$ in the whole document
	6	$\sum_{t_i \in q} IDF(t_i)$ in body
	7	$\sum_{t_i \in q} IDF(t_i)$ in anchor
	8	$\sum_{t_i \in q} IDF(t_i)$ in title
	9	$\sum_{t_i \in q} IDF(t_i)$ in URL
	10	$\sum_{t_i \in q} IDF(t_i)$ in the whole document
	11	$\sum_{t_i \in q \cap d} TF(t_i, d) \cdot IDF(t_i)$ in body
	12	$\sum_{t_i \in q \cap d} TF(t_i, d) \cdot IDF(t_i)$ in anchor
	13	$\sum_{t_i \in q \cap d} TF(t_i, d) \cdot IDF(t_i)$ in title
	14	$\sum_{t_i \in q \cap d} TF(t_i, d) \cdot IDF(t_i)$ in URL
	15	$\sum_{t_i \in q \cap d} TF(t_i, d) \cdot IDF(t_i)$ in the whole document
	16	$LEN(d)$ of body
	17	$LEN(d)$ of anchor
	18	$LEN(d)$ of title
	19	$LEN(d)$ of URL
	20	$LEN(d)$ of the whole document
	21	BM25 of body
	22	BM25 of anchor
	23	BM25 of title
	24	BM25 of URL
	25	BM25 of the whole document
	26	LMIR.ABS of body
	27	LMIR.ABS of anchor
	28	LMIR.ABS of title
	29	LMIR.ABS of URL
	30	LMIR.ABS of the whole document
	31	LMIR.DIR of body
	32	LMIR.DIR of anchor
	33	LMIR.DIR of title
	34	LMIR.DIR of URL
	35	LMIR.DIR of the whole document
	36	LMIR.JM of body
	37	LMIR.JM of anchor
	38	LMIR.JM of title
	39	LMIR.JM of URL

Table 10.2 (Continued)

ID	Feature description
40	LMIR.JM of the whole document
41	Sitemap based term propagation
42	Sitemap based score propagation
43	Hyperlink based score propagation: weighted in-link
44	Hyperlink based score propagation: weighted out-link
45	Hyperlink based score propagation: uniform out-link
46	Hyperlink based feature propagation: weighted in-link
47	Hyperlink based feature propagation: weighted out-link
48	Hyperlink based feature propagation: uniform out-link
49	HITS authority
50	HITS hub
51	PageRank
52	HostRank
53	Topical PageRank
54	Topical HITS authority
55	Topical HITS hub
56	Inlink number
57	Outlink number
58	Number of slash in URL
59	Length of URL
60	Number of child page
61	BM25 of extracted title
62	LMIR.ABS of extracted title
63	LMIR.DIR of extracted title
64	LMIR.JM of extracted title

- Raw information of the documents associated with each query, such as the term frequency and the document length.
- Relational information, such as the hyperlink graph, the sitemap information, and the similarity relationship matrix of the corpus.

With the meta information, one can reproduce existing features, tune their parameters, investigate new features, and perform some advanced research such as relational ranking [14, 15].

10.6 Learning Tasks

The major learning task supported by LETOR is supervised ranking. That is, given a training set that is fully labeled, a learning-to-rank algorithm is employed to learn

Table 10.3 Features for the OHSUMED corpus

ID	Feature description
1	$\sum_{t_i \in q \cap d} TF(t_i, d)$ in title
2	$\sum_{t_i \in q \cap d} \log(TF(t_i, d) + 1)$ in title
3	$\sum_{t_i \in q \cap d} \frac{TF(t_i, d)}{LEN(d)}$ in title
4	$\sum_{t_i \in q \cap d} \log(\frac{TF(t_i, d)}{LEN(d)} + 1)$ in title
5	$\sum_{t_i \in q} \log(|C| \cdot IDF(t_i))$ in title
6	$\sum_{t_i \in q} \log(\log(|C| \cdot IDF(t_i)))$ in title
7	$\sum_{t_i \in q} \log(\frac{|C|}{TF(t_i, C)} + 1)$ in title
8	$\sum_{t_i \in q \cap d} \log(\frac{TF(t_i, d)}{LEN(d)} \cdot \log(|C| \cdot IDF(t_i)) + 1)$ in title
9	$\sum_{t_i \in q \cap d} TF(t_i, d) \cdot \log(|C| \cdot IDF(t_i))$ in title
10	$\sum_{t_i \in q \cap d} \log(\frac{TF(t_i, d)}{LEN(d)} \cdot \frac{|C|}{TF(t_i, C)} + 1)$ in title
11	BM25 of title
12	log(BM25) of title
13	LMIR.DIR of title
14	LMIR.JM of title
15	LMIR.ABS of title
16	$\sum_{t_i \in q \cap d} TF(t_i, d)$ in abstract
17	$\sum_{t_i \in q \cap d} \log(TF(t_i, d) + 1)$ in abstract
18	$\sum_{t_i \in q \cap d} \frac{TF(t_i, d)}{LEN(d)}$ in abstract
19	$\sum_{t_i \in q \cap d} \log(\frac{TF(t_i, d)}{LEN(d)} + 1)$ in abstract
20	$\sum_{t_i \in q} \log(|C| \cdot IDF(t_i))$ in abstract
21	$\sum_{t_i \in q} \log(\log(|C| \cdot IDF(t_i)))$ in abstract
22	$\sum_{t_i \in q} \log(\frac{|C|}{TF(t_i, C)} + 1)$ in abstract
23	$\sum_{t_i \in q \cap d} \log(\frac{TF(t_i, d)}{LEN(d)} \cdot \log(|C| \cdot IDF(t_i)) + 1)$ in abstract
24	$\sum_{t_i \in q \cap d} TF(t_i, d) \cdot \log(|C| \cdot IDF(t_i))$ in abstract
25	$\sum_{t_i \in q \cap d} \log(\frac{TF(t_i, d)}{LEN(d)} \cdot \frac{|C|}{TF(t_i, C)} + 1)$ in abstract
26	BM25 of abstract
27	log(BM25) of abstract
28	LMIR.DIR of abstract
29	LMIR.JM of abstract
30	LMIR.ABS of abstract
31	$\sum_{t_i \in q \cap d} TF(t_i, d)$ in 'title + abstract'
32	$\sum_{t_i \in q \cap d} \log(TF(t_i, d) + 1)$ in 'title + abstract'
33	$\sum_{t_i \in q \cap d} \frac{TF(t_i, d)}{LEN(d)}$ in 'title + abstract'
34	$\sum_{t_i \in q \cap d} \log(\frac{TF(t_i, d)}{LEN(d)} + 1)$ in 'title + abstract'
35	$\sum_{t_i \in q} \log(|C| \cdot IDF(t_i))$ in 'title + abstract'
36	$\sum_{t_i \in q} \log(\log(|C| \cdot IDF(t_i)))$ in 'title + abstract'
37	$\sum_{t_i \in q} \log(\frac{|C|}{TF(t_i, C)} + 1)$ in 'title + abstract'
38	$\sum_{t_i \in q \cap d} \log(\frac{TF(t_i, d)}{LEN(d)} \cdot \log(|C| \cdot IDF(t_i)) + 1)$ in 'title + abstract'

Table 10.3   (Continued)

ID	Feature description
39	$\sum_{t_i \in q \cap d} TF(t_i, d) \cdot \log(|C| \cdot IDF(t_i))$ in 'title + abstract'
40	$\sum_{t_i \in q \cap d} \log(\frac{TF(t_i,d)}{LEN(d)} \cdot \frac{|C|}{TF(qt_i,C)} + 1)$ in 'title + abstract'
41	BM25 of 'title + abstract'
42	log(BM25) of 'title + abstract'
43	LMIR.DIR of 'title + abstract'
44	LMIR.JM of in 'title + abstract'
45	LMIR.ABS of in 'title + abstract'

ranking models. These models are selected using the validation set and finally evaluated on the test set.

In order to make the evaluation more comprehensive, five-fold cross validation
is suggested in LETOR. In particular, each dataset in LETOR is partitioned into five
parts with about the same number of queries, denoted as S1, S2, S3, S4, and S5, in
order to conduct five-fold cross validation. For each fold, three parts are used for
training the ranking model, one part for tuning the hyper parameters of the ranking
algorithm (e.g., the number of iterations in RankBoost [3] and the combination coefficient in the objective function of Ranking SVM [4, 7]), and the remaining part
for evaluating the ranking performance of the learned model (see Table 10.5). The
average performance over the five folds is used to measure the overall performance
of a learning-to-rank algorithm.

One may have noticed that the natural labels in all the LETOR datasets are relevance degrees. As aforementioned, sometimes, pairwise preference and even total
order of the documents are also valid labels. To facilitate learning with such kinds
of labels, in LETOR 4.0, the total order of the labeled documents in MQ2007 and
MQ2008 are derived by heuristics, and used for training.

In addition to the standard supervised ranking, LETOR also supports semi-
supervised ranking and rank aggregation. Different from the task of supervised ranking, semi-supervised ranking considers both judged and unjudged query–document
pairs for training. For rank aggregation, a query is associated with a set of input
ranked lists but not the features for individual documents. The task is to output a
better ranked list by aggregating the multiple input lists.

The LETOR datasets, containing the aforementioned feature representations of
documents, their relevance judgments with respective to queries, and the partitioned
training, validation, and test sets can be downloaded from the official LETOR website, http://research.microsoft.com/~LETOR/.[5]

---

[5]Note that the LETOR datasets are being frequently updated. It is expected that more datasets will
be added in the future. Furthermore, the LETOR website has evolved to be a portal for the research
on learning to rank, which is not limited to the data release only. One can find representative papers,
tutorials, events, research groups, etc. in the area of learning to rank from the website.

Table 10.4 Features for the "Gov2" corpus

ID	Feature description
1	$\sum_{t_i \in q \cap d} TF(t_i, d)$ in body
2	$\sum_{t_i \in q \cap d} TF(t_i, d)$ in anchor
3	$\sum_{t_i \in q \cap d} TF(t_i, d)$ in title
4	$\sum_{t_i \in q \cap d} TF(t_i, d)$ in URL
5	$\sum_{t_i \in q \cap d} TF(t_i, d)$ in the whole document
6	$\sum_{t_i \in q} IDF(t_i)$ in body
7	$\sum_{t_i \in q} IDF(t_i)$ in anchor
8	$\sum_{t_i \in q} IDF(t_i)$ in title
9	$\sum_{t_i \in q} IDF(t_i)$ in URL
10	$\sum_{t_i \in q} IDF(t_i)$ in the whole document
11	$\sum_{t_i \in q \cap d} TF(t_i, d) \cdot IDF(t_i)$ in body
12	$\sum_{t_i \in q \cap d} TF(t_i, d) \cdot IDF(t_i)$ in anchor
13	$\sum_{t_i \in q \cap d} TF(t_i, d) \cdot IDF(t_i)$ in title
14	$\sum_{t_i \in q \cap d} TF(t_i, d) \cdot IDF(t_i)$ in URL
15	$\sum_{t_i \in q \cap d} TF(t_i, d) \cdot IDF(t_i)$ in the whole document
16	$LEN(d)$ of body
17	$LEN(d)$ of anchor
18	$LEN(d)$ of title
19	$LEN(d)$ of URL
20	$LEN(d)$ of the whole document
21	BM25 of body
22	BM25 of anchor
23	BM25 of title
24	BM25 of URL
25	BM25 of the whole document
26	LMIR.ABS of body
27	LMIR.ABS of anchor
28	LMIR.ABS of title
29	LMIR.ABS of URL
30	LMIR.ABS of the whole document
31	LMIR.DIR of body
32	LMIR.DIR of anchor
33	LMIR.DIR of title
34	LMIR.DIR of URL
35	LMIR.DIR of the whole document
36	LMIR.JM of body
37	LMIR.JM of anchor
38	LMIR.JM of title
39	LMIR.JM of URL
40	LMIR.JM of the whole document
41	PageRank
42	Inlink number
43	Outlink number
44	Number of slash in URL
45	Length of URL
46	Number of child page

Table 10.5 Data partitioning
for five-fold cross validation

Folds	Training set	Validation set	Test set
Fold1	{S1, S2, S3}	S4	S5
Fold2	{S2, S3, S4}	S5	S1
Fold3	{S3, S4, S5}	S1	S2
Fold4	{S4, S5, S1}	S2	S3
Fold5	{S5, S1, S2}	S3	S4

10.7 Discussions

LETOR has been widely used in the research community of learning to rank. However, its current version also has limitations. Here we list some of them.

- *Document sampling strategy.* For the datasets based on the "Gov" corpus, the retrieval problem is essentially cast as a re-ranking task (for the top 1000 documents) in LETOR. On one hand, this is a common practice for real-world Web search engines. Usually two rankers are used by a search engine for sake of efficiency: firstly a simple ranker (e.g., BM25 [16]) is used to select some candidate documents, and then a more complex ranker (e.g., the learning-to-rank algorithms as introduced in the book) is used to produce the final ranking result. On the other hand, however, there are also some retrieval applications that should not be cast as a re-ranking task. It would be good to add datasets beyond re-ranking settings to LETOR in the future.
- *Features.* In both academic and industrial communities, more and more features have been studied and applied to improve ranking accuracy. The feature list provided in LETOR is far away from comprehensive. For example, document features (such as document length) are not included in the OHSUMED dataset, and proximity features [18] are not included in all the datasets. It would be helpful to add more features into the LETOR datasets in the future.
- *Scale and diversity of datasets.* As compared with Web search, the scales (number of queries) of the datasets in LETOR are much smaller. To verify the performances of learning-to-rank techniques for real Web search, large scale datasets are needed. Furthermore, although there are nine query sets, there are only three document corpora involved. It would be better to create new datasets using more document corpora in the future.

References

1. Aslam, J.A., Kanoulas, E., Pavlu, V., Savev, S., Yilmaz, E.: Document selection methodologies for efficient and effective learning-to-rank. In: Proceedings of the 32nd Annual International ACM SIGIR Conference on Research and Development in Information Retrieval (SIGIR 2009), pp. 468–475 (2009)
2. Craswell, N., Hawking, D., Wilkinson, R., Wu, M.: Overview of the trec 2003 Web track. In: Proceedings of the 12th Text Retrieval Conference (TREC 2003), pp. 78–92 (2003)

3. Freund, Y., Iyer, R., Schapire, R., Singer, Y.: An efficient boosting algorithm for combining preferences. Journal of Machine Learning Research **4**, 933–969 (2003)
4. Herbrich, R., Obermayer, K., Graepel, T.: Large margin rank boundaries for ordinal regression. In: Advances in Large Margin Classifiers, pp. 115–132 (2000)
5. Hersh, W., Buckley, C., Leone, T.J., Hickam, D.: Ohsumed: an interactive retrieval evaluation and new large test collection for research. In: Proceedings of the 17th Annual International ACM SIGIR Conference on Research and Development in Information Retrieval (SIGIR 1994), pp. 192–201 (1994)
6. Hu, Y., Xin, G., Song, R., Hu, G., Shi, S., Cao, Y., Li, H.: Title extraction from bodies of html documents and its application to web page retrieval. In: Proceedings of the 28th Annual International ACM SIGIR Conference on Research and Development in Information Retrieval (SIGIR 2005), pp. 250–257 (2005)
7. Joachims, T.: Optimizing search engines using clickthrough data. In: Proceedings of the 8th ACM SIGKDD International Conference on Knowledge Discovery and Data Mining (KDD 2002), pp. 133–142 (2002)
8. Liu, T.Y., Xu, J., Qin, T., Xiong, W.Y., Li, H.: LETOR: benchmark dataset for research on learning to rank for information retrieval. In: SIGIR 2007 Workshop on Learning to Rank for Information Retrieval (LR4IR 2007) (2007)
9. Minka, T., Robertson, S.: Selection bias in the LETOR datasets. In: SIGIR 2008 Workshop on Learning to Rank for Information Retrieval (LR4IR 2008) (2008)
10. Nie, L., Davison, B.D., Qi, X.: Topical link analysis for web search. In: Proceedings of the 29th Annual International ACM SIGIR Conference on Research and Development in Information Retrieval (SIGIR 2006), pp. 91–98 (2006)
11. Qin, T., Liu, T.Y., Lai, W., Zhang, X.D., Wang, D.S., Li, H.: Ranking with multiple hyperplanes. In: Proceedings of the 30th Annual International ACM SIGIR Conference on Research and Development in Information Retrieval (SIGIR 2007), pp. 279–286 (2007)
12. Qin, T., Liu, T.Y., Xu, J., Li, H.: How to make LETOR more useful and reliable. In: SIGIR 2008 Workshop on Learning to Rank for Information Retrieval (LR4IR 2008) (2008)
13. Qin, T., Liu, T.Y., Zhang, X.D., Chen, Z., Ma, W.Y.: A study of relevance propagation for web search. In: Proceedings of the 28th Annual International ACM SIGIR Conference on Research and Development in Information Retrieval (SIGIR 2005), pp. 408–415 (2005)
14. Qin, T., Liu, T.Y., Zhang, X.D., Wang, D., Li, H.: Learning to rank relational objects and its application to web search. In: Proceedings of the 17th International Conference on World Wide Web (WWW 2008), pp. 407–416 (2008)
15. Qin, T., Liu, T.Y., Zhang, X.D., Wang, D.S., Li, H.: Global ranking using continuous conditional random fields. In: Advances in Neural Information Processing Systems 21 (NIPS 2008), pp. 1281–1288 (2009)
16. Robertson, S.E.: Overview of the okapi projects. Journal of Documentation **53**(1), 3–7 (1997)
17. Shakery, A., Zhai, C.: A probabilistic relevance propagation model for hypertext retrieval. In: Proceedings of the 15th International Conference on Information and Knowledge Management (CIKM 2006), pp. 550–558 (2006)
18. Tao, T., Zhai, C.: An exploration of proximity measures in information retrieval. In: Proceedings of the 30th Annual International ACM SIGIR Conference on Research and Development in Information Retrieval (SIGIR 2007), pp. 295–302 (2007)
19. Xu, J., Li, H.: Adarank: a boosting algorithm for information retrieval. In: Proceedings of the 30th Annual International ACM SIGIR Conference on Research and Development in Information Retrieval (SIGIR 2007), pp. 391–398 (2007)
20. Xue, G.R., Yang, Q., Zeng, H.J., Yu, Y., Chen, Z.: Exploiting the hierarchical structure for link analysis. In: Proceedings of the 28th Annual International ACM SIGIR Conference on Research and Development in Information Retrieval (SIGIR 2005), pp. 186–193 (2005)

Chapter 11
Experimental Results on LETOR

Abstract In this chapter, we take the official evaluation results published at the LETOR website as the source to perform discussions on the performances of different learning-to-rank methods.

11.1 Experimental Settings

Three widely used measures are adopted for the evaluation in the LETOR datasets: P@k [1], MAP [1], and NDCG@k [6]. For a given ranking model, the evaluation results in terms of these three measures can be computed by the official evaluation tool provided in LETOR.

LETOR official baselines include several learning-to-rank algorithms, such as linear regression, belonging to the pointwise approach; Ranking SVM [5, 7], RankBoost [4], and FRank [8], belonging to the pairwise approach; ListNet [2], AdaRank [10], and SVM^{map} [11], belonging to the listwise approach. To make fair comparisons, the same setting for all the algorithms are adopted. Firstly, most algorithms use the linear scoring function, except RankBoost and FRank, which uses binary weak rankers. Secondly, all the algorithms use MAP on the validation set for model selection. Some detailed experimental settings are listed here.

- As for linear regression, the validation set is used to select a good mapping from the ground-truth labels to real values.
- For Ranking SVM, the public tool of SVMlight is employed and the validation set is used to tune the parameter λ in its loss function.
- For RankBoost, the weak ranker is defined on the basis of a single feature with 255 possible thresholds. The validation set is used to determine the best number of iterations.
- For FRank, the validation set is used to determine the number of weak learners in the generalized additive model.

Note that there have been several other empirical studies [9, 12] in the literature, based on LETOR and other datasets. The conclusions drawn from these studies are similar to what we will introduce in this chapter.

T.-Y. Liu, *Learning to Rank for Information Retrieval*,
DOI 10.1007/978-3-642-14267-3_11, © Springer-Verlag Berlin Heidelberg 2011

Table 11.1 Results on the TD2003 dataset

Algorithm	NDCG@1	NDCG@3	NDCG@10	P@1	P@3	P@10	MAP
Regression	0.320	0.307	0.326	0.320	0.260	0.178	0.241
RankSVM	0.320	0.344	0.346	0.320	0.293	0.188	0.263
RankBoost	0.280	0.325	0.312	0.280	0.280	0.170	0.227
FRank	0.300	0.267	0.269	0.300	0.233	0.152	0.203
ListNet	0.400	0.337	0.348	0.400	0.293	0.200	0.275
AdaRank	0.260	0.307	0.306	0.260	0.260	0.158	0.228
SVM^{map}	0.320	0.320	0.328	0.320	0.253	0.170	0.245

- For ListNet, the validation set is used to determine the best mapping from the ground-truth label to scores in order to use the Plackett–Luce model, and to determine the optimal number of iterations in the gradient descent process.
- For AdaRank, MAP is set as the evaluation measure to be optimized, and the validation set is used to determine the number of iterations.
- For SVM^{map}, the publicly available tool SVM^{map} (http://projects.yisongyue.com/svmmap/) is employed, and the validation set is used to determine the parameter λ in its loss function.

11.2 Experimental Results on LETOR 3.0

The ranking performances of the aforementioned algorithms on the LETOR 3.0 datasets are listed in Tables 11.1, 11.2, 11.3, 11.4, 11.5, 11.6, and 11.7. According to these experimental results, we find that the listwise ranking algorithms perform very well on most datasets. Among the three listwise ranking algorithms, ListNet seems to be better than the other two. AdaRank and SVM^{map} obtain similar performances. Pairwise ranking algorithms obtain good ranking accuracy on some (although not all) datasets. For example, RankBoost offers the best performance on TD2004 and NP2003; Ranking SVM shows very promising results on NP2003 and NP2004; and FRank achieves very good results on TD2004 and NP2004. Comparatively speaking, simple linear regression performs worse than the pairwise and listwise ranking algorithms. Its results are not so good on most datasets.

We have also observed that most ranking algorithms perform differently on different datasets. They may perform very well on some datasets, but not so well on the others. To evaluate the overall ranking performances of an algorithm, we use the number of other algorithms that it can beat over all the seven datasets as a measure. That is,

$$S_i(M) = \sum_{j=1}^{7} \sum_{k=1}^{7} I_{\{M_i(j) > M_k(j)\}}$$

where j is the index of a dataset, i and k are the indexes of algorithms, $M_i(j)$ is the performance of the ith algorithm on the jth dataset, and $I_{\{\cdot\}}$ is the indicator function.

Table 11.2 Results on the TD2004 dataset

Algorithm	NDCG@1	NDCG@3	NDCG@10	P@1	P@3	P@10	MAP
Regression	0.360	0.335	0.303	0.360	0.333	0.249	0.208
RankSVM	0.413	0.347	0.307	0.413	0.347	0.252	0.224
RankBoost	0.507	0.430	0.350	0.507	0.427	0.275	0.261
FRank	0.493	0.388	0.333	0.493	0.378	0.262	0.239
ListNet	0.360	0.357	0.317	0.360	0.360	0.256	0.223
AdaRank	0.413	0.376	0.328	0.413	0.369	0.249	0.219
SVM^{map}	0.293	0.304	0.291	0.293	0.302	0.247	0.205

Table 11.3 Results on the NP2003 dataset

Algorithm	NDCG@1	NDCG@3	NDCG@10	P@1	P@3	P@10	MAP
Regression	0.447	0.614	0.665	0.447	0.220	0.081	0.564
RankSVM	0.580	0.765	0.800	0.580	0.271	0.092	0.696
RankBoost	0.600	0.764	0.807	0.600	0.269	0.094	0.707
FRank	0.540	0.726	0.776	0.540	0.253	0.090	0.664
ListNet	0.567	0.758	0.801	0.567	0.267	0.092	0.690
AdaRank	0.580	0.729	0.764	0.580	0.251	0.086	0.678
SVM^{map}	0.560	0.767	0.798	0.560	0.269	0.089	0.687

Table 11.4 Results on the NP2004 dataset

Algorithm	NDCG@1	NDCG@3	NDCG@10	P@1	P@3	P@10	MAP
Regression	0.373	0.555	0.653	0.373	0.200	0.082	0.514
RankSVM	0.507	0.750	0.806	0.507	0.262	0.093	0.659
RankBoost	0.427	0.627	0.691	0.427	0.231	0.088	0.564
FRank	0.480	0.643	0.729	0.480	0.236	0.093	0.601
ListNet	0.533	0.759	0.812	0.533	0.267	0.094	0.672
AdaRank	0.480	0.698	0.749	0.480	0.244	0.088	0.622
SVM^{map}	0.520	0.749	0.808	0.520	0.267	0.096	0.662

It is clear that the larger $S_i(M)$ is, the better the ith algorithm performs. For ease of reference, we refer to this measure as the *winning number*. Table 11.8 shows the winning number for all the algorithms under investigation. From this table, we have the following observations.

1. In terms of NDCG@1, P@1 and P@3, the listwise ranking algorithms perform the best, followed by the pairwise ranking algorithms, while the pointwise ranking algorithm performs the worst. Among the three listwise ranking algorithms, ListNet is better than AdaRank and SVM^{map}. The three pairwise ranking al-

Table 11.5 Results on the HP2003 dataset

Algorithm	NDCG@1	NDCG@3	NDCG@10	P@1	P@3	P@10	MAP
Regression	0.420	0.510	0.594	0.420	0.211	0.088	0.497
RankSVM	0.693	0.775	0.807	0.693	0.309	0.104	0.741
RankBoost	0.667	0.792	0.817	0.667	0.311	0.105	0.733
FRank	0.653	0.743	0.797	0.653	0.289	0.106	0.710
ListNet	0.720	0.813	0.837	0.720	0.320	0.106	0.766
AdaRank	0.733	0.805	0.838	0.733	0.309	0.106	0.771
SVM^{map}	0.713	0.779	0.799	0.713	0.309	0.100	0.742

Table 11.6 Results on the HP2004 dataset

Algorithm	NDCG@1	NDCG@3	NDCG@10	P@1	P@3	P@10	MAP
Regression	0.387	0.575	0.646	0.387	0.213	0.08	0.526
RankSVM	0.573	0.715	0.768	0.573	0.267	0.096	0.668
RankBoost	0.507	0.699	0.743	0.507	0.253	0.092	0.625
FRank	0.600	0.729	0.761	0.600	0.262	0.089	0.682
ListNet	0.600	0.721	0.784	0.600	0.271	0.098	0.690
AdaRank	0.613	0.816	0.832	0.613	0.298	0.094	0.722
SVM^{map}	0.627	0.754	0.806	0.627	0.280	0.096	0.718

Table 11.7 Results on the OHSUMED dataset

Algorithm	NDCG@1	NDCG@3	NDCG@10	P@1	P@3	P@10	MAP
Regression	0.446	0.443	0.411	0.597	0.577	0.466	0.422
RankSVM	0.496	0.421	0.414	0.597	0.543	0.486	0.433
RankBoost	0.463	0.456	0.430	0.558	0.561	0.497	0.441
FRank	0.530	0.481	0.443	0.643	0.593	0.501	0.444
ListNet	0.533	0.473	0.441	0.652	0.602	0.497	0.446
AdaRank	0.539	0.468	0.442	0.634	0.590	0.497	0.449
SVM^{map}	0.523	0.466	0.432	0.643	0.580	0.491	0.445

gorithms achieve comparable results, among which Ranking SVM seems to be slightly better than the other two algorithms.

2. In terms of NDCG@3 and NDCG@10, ListNet and AdaRank perform much better than the pairwise and pointwise ranking algorithms, while the performance of SVM^{map} is very similar to the pairwise ranking algorithms.

3. In terms of P@10, ListNet performs much better than the pairwise and pointwise ranking algorithms, while the performances of AdaRank and SVM^{map} are not so good as those of the pairwise ranking algorithms.

Table 11.8   Winning number of each algorithm

Algorithm	NDCG@1	NDCG@3	NDCG@10	P@1	P@3	P@10	MAP
Regression	4	4	4	5	5	5	4
RankSVM	21	22	22	21	22	22	24
RankBoost	18	22	22	17	22	23	19
FRank	18	19	18	18	17	23	15
ListNet	29	31	33	30	32	35	33
AdaRank	26	25	26	23	22	16	27
SVM^{map}	23	24	22	25	20	17	25

Table 11.9   Results on the MQ2007 dataset

Algorithm	NDCG@1	NDCG@3	NDCG@10	P@1	P@3	P@10	MAP
RankSVM	0.410	0.406	0.444	0.475	0.432	0.383	0.464
RankBoost	0.413	0.407	0.446	0.482	0.435	0.386	0.466
ListNet	0.400	0.409	0.444	0.464	0.433	0.380	0.465
AdaRank	0.382	0.398	0.434	0.439	0.423	0.374	0.458

4. In terms of MAP, the listwise ranking algorithms are in general better than the pairwise ranking algorithms. Furthermore, the variance among the three pairwise ranking algorithms in terms of MAP is much larger than that in terms of other measures (e.g., P@1, 3, and 10). The possible explanation is that since MAP involves all the documents associated with a query in the evaluation process, it can better differentiate algorithms.

To summarize, the experimental results indicate that the listwise algorithms have certain advantages over other algorithms, especially for the top positions of the ranking result.

11.3   Experimental Results on LETOR 4.0

As for the datasets in LETOR 4.0, not many baselines have been tested. The only available results are about Ranking SVM, RankBoost, ListNet, and AdaRank, as shown in Tables 11.9 and 11.10.

From the experimental results, we can find that the differences between the algorithms under investigation are not very large. For example, Ranking SVM, RankBoost, and ListNet perform similarly and a little bit better than AdaRank on MQ2007. RankBoost, ListNet, and AdaRank perform similarly on MQ 2008, and Ranking SVM performs a little bit worse. These observations are not quite in accordance with the results obtained on LETOR 3.0.

The possible explanation of the above experimental results is as follows. As we know, the datasets in LETOR 4.0 have more queries than the datasets in LETOR 3.0;

Table 11.10 Results on the MQ2008 dataset

Algorithm	NDCG@1	NDCG@3	NDCG@10	P@1	P@3	P@10	MAP
RankSVM	0.363	0.429	0.228	0.427	0.390	0.249	0.470
RankBoost	0.386	0.429	0.226	0.458	0.392	0.249	0.478
ListNet	0.375	0.432	0.230	0.445	0.384	0.248	0.478
AdaRank	0.375	0.437	0.230	0.443	0.390	0.245	0.476

however, the number of features in LETOR 4.0 is smaller than that in LETOR 3.0. In this case, we hypothesize that the large scale of the dataset has enabled all the learning-to-rank algorithms to (almost) fully realize the potential in the current feature representation. In other words, the datasets have become saturated and cannot well distinguish different learning algorithms. A similar phenomenon has been observed in some other work [3], especially when the scale of the training data is large.

In this regard, it is sometimes not enough to increase the size of the datasets in order to get meaningful evaluation results on learning-to-rank algorithms. Enriching the feature representation is also very important, if not more important. This should be a critical future work for the research community of learning to rank.

11.4 Discussions

Here we would like to point out that the above experimental results are still primal, since the LETOR baselines have not been fine-tuned and the performance of almost every baseline algorithm can be further improved.

- Most baselines in LETOR use linear scoring functions. For such a complex problem as ranking, linear scoring functions may be too simple. From the experimental results, we can see that the performances of these algorithms are still much lower than the perfect ranker (whose MAP and NDCG are both one). This partially verifies the limited power of linear scoring function. Furthermore, query features such as inverted document frequency cannot be effectively used by linear ranking functions. In this regard, if we use non-linear ranking functions, we should be able to greatly improve the performances of the baseline algorithms.
- As for the loss functions in the baseline algorithms, we also have a large space to make further improvement. For example, in regression, we can add a regularization term to make it more robust (actually according to the experiments conducted in [3], the regularized linear regression model performs much better than the original linear regression model, and its ranking performance is comparable to many pairwise ranking methods); for ListNet, we can also add a regularization term to its loss function and make it more generalizable to the test set.

11.5 Exercises

11.1 Implement all the learning-to-rank algorithms introduced in previous chapters and test them on the LETOR datasets.

11.2 Study the contribution of each feature to the ranking performance. Suppose the linear scoring function is used in the learning-to-rank algorithms, please study the differences in the weights of the features learned by different algorithms, and explain the differences.

11.3 Perform case studies to explain the experimental results observed in this chapter.

11.4 Analyze the computational complexity of each learning-to-rank algorithm.

11.5 Select one learning-to-rank algorithm and perform training for linear and non-linear ranking models respectively. Compare the ranking performances of two models and discuss their advantages and disadvantages.

11.6 Study the change of the ranking performance with respect to different evaluation measures. For example, with the increasing k value, how do the ranking performances of different algorithms compare with each other in terms of NDCG@k?

References

1. Baeza-Yates, R., Ribeiro-Neto, B.: Modern Information Retrieval. Addison-Wesley, Reading (1999)
2. Cao, Z., Qin, T., Liu, T.Y., Tsai, M.F., Li, H.: Learning to rank: from pairwise approach to listwise approach. In: Proceedings of the 24th International Conference on Machine Learning (ICML 2007), pp. 129–136 (2007)
3. Chapelle, O.: Direct optimization for ranking. In: Keynote Speech, SIGIR 2009 Workshop on Learning to Rank for Information Retrieval (LR4IR 2009) (2009)
4. Freund, Y., Iyer, R., Schapire, R., Singer, Y.: An efficient boosting algorithm for combining preferences. Journal of Machine Learning Research **4**, 933–969 (2003)
5. Herbrich, R., Obermayer, K., Graepel, T.: Large margin rank boundaries for ordinal regression. In: Advances in Large Margin Classifiers, pp. 115–132 (2000)
6. Järvelin, K., Kekäläinen, J.: Cumulated gain-based evaluation of IR techniques. ACM Transactions on Information Systems **20**(4), 422–446 (2002)
7. Joachims, T.: Optimizing search engines using clickthrough data. In: Proceedings of the 8th ACM SIGKDD International Conference on Knowledge Discovery and Data Mining (KDD 2002), pp. 133–142 (2002)
8. Tsai, M.F., Liu, T.Y., Qin, T., Chen, H.H., Ma, W.Y.: Frank: a ranking method with fidelity loss. In: Proceedings of the 30th Annual International ACM SIGIR Conference on Research and Development in Information Retrieval (SIGIR 2007), pp. 383–390 (2007)
9. Verberne, S., Halteren, H.V., Theijssen, D., Raaijmakers, S., Boves, L.: Learning to rank qa data. In: SIGIR 2009 Workshop on Learning to Rank for Information Retrieval (LR4IR 2009) (2009)
10. Xu, J., Li, H.: Adarank: a boosting algorithm for information retrieval. In: Proceedings of the 30th Annual International ACM SIGIR Conference on Research and Development in Information Retrieval (SIGIR 2007), pp. 391–398 (2007)

11. Yue, Y., Finley, T., Radlinski, F., Joachims, T.: A support vector method for optimizing average precision. In: Proceedings of the 30th Annual International ACM SIGIR Conference on Research and Development in Information Retrieval (SIGIR 2007), pp. 271–278 (2007)

12. Zhang, M., Kuang, D., Hua, G., Liu, Y., Ma, S.: Is learning to rank effective for web search. In: SIGIR 2009 Workshop on Learning to Rank for Information Retrieval (LR4IR 2009) (2009)

Chapter 12
Other Datasets

Abstract In this chapter, we introduce two new benchmark datasets, released by Yahoo and Microsoft. These datasets originate from the training data used in commercial search engines and are much larger than the LETOR datasets in terms of both number of queries and number of documents per query.

12.1 Yahoo! Learning-to-Rank Challenge Datasets

Yahoo! Labs organized a learning-to-rank challenge from March 1 to May 31, 2010. Given that learning to rank has become a very hot research area, many researchers have participated in this challenge and tested their own algorithms. There were 4,736 submissions coming from 1,055 teams, and the results of this challenge were summarized at a workshop at the 27th International Conference on Machine Learning (ICML 2010) in Haifa, Israel. The official website of this challenge is http://learningtorankchallenge.yahoo.com/.

According to the website of the challenge, the datasets used in this challenge come from web search ranking and are of a subset of what Yahoo! uses to train its ranking function.

There are two datasets for this challenge, each corresponding to a different country: a large one and a small one. The two datasets are related, but also different to some extent. Each dataset is divided into three sets: training, validation, and test. The statistics for the various sets are as shown in Tables 12.1 and 12.2.

The datasets consist of feature vectors extracted from query–URL pairs along with relevance judgments. The relevance judgments can take five different values from 0 (irrelevant) to 4 (perfectly relevant). There are 700 features in total. Some of them are only defined in one dataset, while some others are defined in both sets. When a feature is undefined for a set, its value is 0. All the features have been normalized to be in the [0, 1] range. The queries, URLs, and feature descriptions are not disclosed, only the feature values, because of the following reason. Feature engineering is a critical component of any commercial search engine. For this reason, search engine companies rarely disclose the features they use. Releasing the queries and URLs would lead to a risk of reverse engineering of the features. This is a reasonable consideration; however, it will prevent information retrieval researchers from studying what kinds of feature are the most effective ones for learning-to-rank.

T.-Y. Liu, *Learning to Rank for Information Retrieval*,   153
DOI 10.1007/978-3-642-14267-3_12, © Springer-Verlag Berlin Heidelberg 2011

Table 12.1 Dataset 1 for
Yahoo! learning-to-rank
challenge

	Training	Validation	Test
Number of queries	19,944	2,994	6,983
Number of URLs	473,134	71,083	165,660

Table 12.2 Dataset 2 for
Yahoo! learning-to-rank
challenge

	Training	Validation	Test
Number of queries	1,266	1,266	3,798
Number of URLs	34,815	34,881	103,174

The competition is divided into two tracks:

- A standard learning-to-rank track, using only the larger dataset.
- A transfer learning track, where the goal is to leverage the training set from dataset 1 to build a better ranking function on dataset 2.

Two measures are used for the evaluation of the competition: NDCG [2] and Expected Reciprocal Rank (ERR). The definition of ERR is given as follows:

$$ERR = \sum_{i=1}^{m} \frac{1}{m} \frac{G(y_i)}{16} \prod_{j=1}^{i-1}\left(1 - \frac{G(y_j)}{16}\right), \quad \text{with } G(y) = 2^y - 1. \quad (12.1)$$

The datasets can be downloaded from the sandbox of Yahoo! Research.[1] There are no official baselines on these datasets, however, most of the winners of the competition have published the details of their algorithms in the workshop proceedings. They can serve as meaningful baselines.

12.2 Microsoft Learning-to-Rank Datasets

Microsoft Research Asia released two large-scale datasets for the research on learning to rank in May 2010: MSLR-WEB30k and MSLR-WEB10K. MSLR-WEB30K actually has 31,531 queries and 3,771,126 documents. Up to the writing of this book, the MSLR-WEB30k dataset is the largest publicly-available dataset for the research on learning to rank. MSLR-WEB10K is a random sample of MSLR-WEB30K, which has 10,000 queries and 1,200,193 documents.

In the two datasets, queries and URLs are represented by IDs, with a similar reason to the non-disclosure of queries and URLs in the Yahoo! Learning-to-Rank Challenge datasets. The Microsoft datasets consist of feature vectors extracted from query–URL pairs along with relevance judgments:

---

[1] http://webscope.sandbox.yahoo.com/.

- The relevance judgments are obtained from a retired labeling set of Microsoft Bing search engine, which take five values from 0 (irrelevant) to 4 (perfectly relevant).
- There are in total 136 features. These features are not from Bing but extracted by Microsoft Research. All the features are widely used in the research community, including query–document matching features, document features, Web graph features, and user behavior features. The detailed feature list can be found at the official website of Microsoft Learning-to-Rank Datasets.[2] The availability of the information about these features will enable researchers to study the impact of a feature on the ranking performance.

The measures used in Microsoft Learning-to-Rank Datasets are NDCG, P@k [1], and MAP [1], just as in the LETOR datasets [3]. Furthermore, the datasets have been partitioned into five parts for five-fold cross validation, also with the same strategy as in LETOR. Currently there are no official baselines on these datasets either. Researchers need to implement their own baselines.

12.3 Discussions

Given the above trends of releasing larger datasets, we believe that with the continued efforts from the entire research community as well as the industry, more data resources for learning to rank will be available and the research on learning to rank for information retrieval can be significantly advanced.

References

1. Baeza-Yates, R., Ribeiro-Neto, B.: Modern Information Retrieval. Addison-Wesley, Reading (1999)
2. Järvelin, K., Kekäläinen, J.: Cumulated gain-based evaluation of IR techniques. ACM Transactions on Information Systems 20(4), 422–446 (2002)
3. Liu, T.Y., Xu, J., Qin, T., Xiong, W.Y., Li, H.: LETOR: Benchmark dataset for research on learning to rank for information retrieval. In: SIGIR 2007 Workshop on Learning to Rank for Information Retrieval (LR4IR 2007) (2007)

---

[2]http://research.microsoft.com/~MSLR.

Part V
Practical Issues in Learning to Rank

In this part, we will introduce how to apply the learning-to-rank technologies mentioned in the previous parts to solve real ranking problems. This task turns out not to be as trivial as expected. One needs to deal with training data generation, feature extraction, query and document selection, feature selection, and several other issues. We will review previous works on these aspects, and give several examples of applying learning-to-rank technologies.

After reading this part, the readers are expected to understand the practical issues in using learning-to-rank technologies, and to be able to adopt an existing learning-to-rank algorithm in their own applications.

Chapter 13
Data Preprocessing for Learning to Rank

Abstract This chapter is concerned with data processing for learning to rank. In order to learn an effective ranking model, the first step is to prepare high-quality training data. There are several important issues to be considered regarding the training data. First, it should be considered how to get the data labeled on a large scale but at a low cost. Click-through log mining is one of the feasible approaches for this purpose. Second, since the labeled data are not always correct and effective, selection of the queries and documents, as well as their features should also be considered. In this chapter, we will review several pieces of previous work on these topics, and also make discussions on the future work.

13.1 Overview

In the previous chapters, we have introduced different learning-to-rank methods. Throughout the introduction, we have assumed that the training data (queries, associated documents, and their feature representations) are already available, and have focused more on the algorithmic aspect. However, in practice, how to collect and process the training data are also issues that we need to consider.

The most straightforward approach of obtaining training data is to ask human annotators to label the relevance of a given document with respect to a query. However, in practice, there may be several problems with this approach. First, the human annotation is always costly. Therefore, it is not easy to obtain a large amount of labeled data. As far as we know, the largest labeled dataset used in published papers only contains tens of thousands of queries, and millions of documents. Considering that the query space is almost infinite (users can issue any words or combinations of words as queries, and the query vocabulary is constantly evolving), such a training set might not be sufficient for effective training. In such a case, it is highly demanding if we can find a more cost-effective way to collect useful training data. Second, even if we can afford a certain amount of cost, there is still a tough decision to be made on how to spend these budget. Is it more beneficial to label more queries, or to label more documents per query? Shall we label the data independent of the training process, or only ask human annotators to label those documents that have the biggest contribution to the training process?

T.-Y. Liu, *Learning to Rank for Information Retrieval*,
DOI 10.1007/978-3-642-14267-3_13, © Springer-Verlag Berlin Heidelberg 2011

In this chapter, we will try to answer the above questions. In particular, we will first introduce various models for user click behaviors and discuss how they can be used to automatically mine ground-truth labels for learning to rank. Then we will discuss the problem of data selection for learning to rank, which includes document selection for labeling, and document/feature selection for training.

13.2 Ground Truth Mining from Logs

13.2.1 User Click Models

Most commercial search engines log users' click behaviors during their interaction with the search interface. Such click logs embed important clues about user satisfaction with a search engine and can provide a highly valuable source of relevance information. As compared to human judgment, click information is much cheaper to obtain and can reflect the up-to-date relevance (relevance will change along with time). However, clicks are also known to be biased and noisy. Therefore, it is necessary to develop some models to remove the bias and noises in order to obtain reliable relevance labels.

Classical click models include the position models [10, 13, 28] and the cascade model [10]. A position model assumes that a click depends on both relevance and examination. Each document has a certain probability of being examined, which decays by and only depends on rank positions. A click on a document indicates that the document is examined and considered relevant by the user. However this model treats the individual documents in a search result page independently and fails to capture the interdependency between documents in the examination probability. The cascade model assumes that users examine the results sequentially and stop as soon as a relevant document is clicked. Here, the probability of examination is indirectly determined by two factors: the rank of the document and the relevance of all previous documents. The cascade model makes a strong assumption that there is only one click per search and hence it could not explain the abandoned search or search with multiple clicks.

To sum up, there are at least the following problems with the aforementioned classical models.

- The models cannot effectively deal with multiple clicks in a session.
- The models cannot distinguish perceived relevance and actual relevance. Because users cannot examine the content of a document until they click on the document, the decision to click is made based on perceived relevance. While there is a strong correlation between perceived relevance and actual relevance, there are also many cases where they differ.
- The models cannot naturally lead to a preference probability on a pair of documents, while such preference information is required by many pairwise ranking methods.

Fig. 13.1 The DCM model

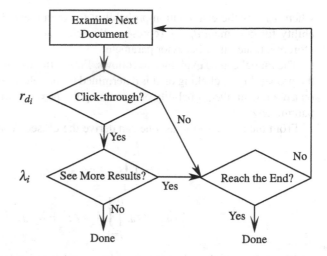

In recent years, several new click models have been proposed in order to solve the aforementioned problems. In this subsection, we will introduce three representative models, namely the Dependent Click Model [19], Bayesian Browsing Model [24], and Dynamic Bayesian Network Click Model [7].

13.2.1.1 Dependent Click Model

As aforementioned, the cascade model assumes that a user abandons examination of web documents upon the first click. This unfortunately restricts the modeling power to query sessions with at most one click, which leaves the gap open for real-world applications where multiple clicks are possible, especially for informational queries.

In order to tackle the problem, in [19], a dependent click model (DCM) is proposed, which generalizes the cascade model to multiple clicks by including a set of position-dependent parameters to model probabilities that an average user returns to the search result page and resumes the examination after a click. The DCM model can be illustrated using Fig. 13.1.

According to Fig. 13.1, a position-dependent parameter λ_i is used to reflect the chance that the user would like to see more results after a click at position i. In case of a skip (no click), the next document is examined with probability one. The λ_i are a set of user behavior parameters shared over multiple query sessions.

The examination and click probabilities in the DCM model can be specified in the following iterative process ($1 \leq i \leq m$, where m is the number of documents in the search result):

$$e_{d_1,1} = 1,$$
$$c_{d_i,i} = e_{d_i,i} r_{d_i},$$
$$e_{d_{i+1},i+1} = \lambda_i c_{d_i,i} + (e_{d_i,i} - c_{d_i,i}),$$

(13.1)

where $e_{d_i,i}$ is the examination probability for document d_i, $c_{d_i,i}$ is the click probability for document d_i, r_{d_i} represents the relevance of document d_i, and λ_i is the aforementioned user behavior parameter.

The model can be explained as follows. First, after the user examines a document, the probability of clicking on it is determined by its relevance. Second, after clicking on a document, the probability of examining the next document is determined by the parameter λ_i.

From the above formulas, one can derive the closed-form equations

$$
\begin{aligned}
e_{d_i,i} &= \prod_{j=1}^{i-1}(1 - r_{d_j} + \lambda_j r_{d_j}), \\
c_{d_i,i} &= r_{d_i} \prod_{j=1}^{i-1}(1 - r_{d_j} + \lambda_j r_{d_j}).
\end{aligned}
\tag{13.2}
$$

Given the actual click events $\{C_i\}$ in a query session and the document impression $\{d_i\}$, a lower bound of its log-likelihood before position k can be obtained as

$$
L_{\text{DCM}} \geq \sum_{i=1}^{k-1}\left(C_i \log r_{d_i} + (1 - C_i)\log(1 - r_{d_i})\right)
$$

$$
+ \sum_{i=1}^{k-1} C_i \log \lambda_i + \log(1 - \lambda_k).
\tag{13.3}
$$

By maximizing the above lower bound of the log-likelihood, one can get the best estimates for the relevance and user behavior parameters:

$$
r_d = \frac{\#\,\text{Clicks on } d}{\text{Impressions of } d \text{ before position } k},
\tag{13.4}
$$

$$
\lambda_i = 1 - \frac{\#\,\text{Query sessions when last clicked position is } i}{\#\,\text{Query sessions when position } i \text{ is clicked}}.
\tag{13.5}
$$

Experimental results on the click-through logs of a commercial search engine have shown that the DCM model can perform effectively. The model is further enhanced in a follow-up work, referred to as the Click Chain Model [18]. The new model removes some unrealistic assumptions in DCM (e.g., users would scan the entire list of the search results), and thus performs better on tail queries and can work in a more efficient manner.

13.2.1.2 Bayesian Browsing Model

As can be seen, both the DCM model and the cascaded model, despite of their different user behavior assumptions, follow the point-estimation philosophy: the estimated relevance for each query–document pair is a single number normalized to

Fig. 13.2 The BBM model

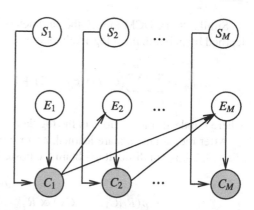

[0,1], and a larger value indicates stronger relevance. While this single-number estimate could suffice for many usages, it nevertheless falls short of capacity for broader
applications. For example, it is unspecified how to estimate the probability that document d_u is preferred to d_v for the same query when their relevances are r_u and r_v
respectively.

Note that many learning-to-rank algorithms, especially the pairwise ranking algorithms, require pairwise preference relationship as the training data. It is desirable
that we can also mine such pairwise information from the click-through data in a
principled way. To tackle this issue, a new click model named Bayesian Browsing
Model (BBM) is proposed in [24]. By virtue of the Bayesian approach to modeling
the document relevance, the preference probability between multiple documents is
well-defined and can be computed based on document relevance posteriors.

The BBM model can be vividly illustrated using Fig. 13.2.

From the figure we can see that the model consists of three layers of random variables. The top layer $S = (S_1, S_2, \ldots, S_M)$ represents nominal relevance variables
with $S_i = R_{\pi^{-1}(i)}$ (here R_j is the relevance of document j, and π is the ranked list
of the search result). The other two layers E and C represent examination and click
events, respectively. The full model specification is as follows:

$$
\begin{aligned}
P(E_1 = 1) &= \beta_{0,1}, \\
P(C_i = 1 | E_i = 0, S_i) &= 0, \\
P(C_i = 1 | E_i = 1, S_i) &= S_i, \\
P(E_i = 1 | C_1, \ldots, C_{i-1}) &= \beta_{t_i, s_i},
\end{aligned}
\tag{13.6}
$$

where $t_i = \arg\max_{l < t}(C_l = 1)$, and $s_i = i - t_i$.

Detailed explanation of the model is given as follows. First, if a document is
not examined by the user, it will not be clicked no matter whether it is relevant or
not. Second, if a document has been examined by the user, the probability that it
is clicked is determined by the relevance of the document. Third, given previous
clicks, the probability of examining the next document is parameterized by β_{t_i, s_i},
which is dependent on the preceding click position t_i and the distance from the
current position to the preceding click s_i.

Similar to the DCM model, the parameters in the above BBM model can also be learned by maximizing a log-likelihood of the observed clicks:

$$L_{BBM} = \sum_{k=1}^{m} \sum_{i=1} \left(C_i^k \log(\beta_{t_i^k, s_i^k}/2) + \left(1 - C_i^k\right) \log(1 - \beta_{t_i^k, s_i^k}/2) \right), \qquad (13.7)$$

where k indexes the sessions in the log data.

After the parameters are learned, according to the graphical model as shown in Fig. 13.2, it is not difficult to obtain the posterior probability of variables R_j:

$$p(R_i | C_1, \ldots, C_N) \propto R_j^{e_0} \prod_{j=1}^{m(m+1)/2} (1 - \gamma_j R_i)^{e_j}, \qquad (13.8)$$

where

$$\gamma_{\frac{t(2m-t+1)}{2}+s} = \beta_{t,s}$$

$$e_0 = \sum_{k=1}^{m} \sum_{i=1}^{m} I_{\{(C_i^k=1) \wedge (\pi^{-1}(i)=j)\}} \qquad (13.9)$$

$$e_{\frac{t(2m-t+1)}{2}+s} = \sum_{k=1}^{m} \sum_{i=1}^{m} I_{\{(C_i^k=0) \wedge (\pi^{-1}(i)=j) \wedge (t_i^k=t) \wedge (s_i^k=s)\}}.$$

With this probability, one can easily get the pairwise preference probability:

$$p(d_j \succ d_i) = P(R_j > R_i) = \int \int_{R_j > R_i} p_j(R_j) p_i(R_i) \, dR_j \, dR_i. \qquad (13.10)$$

This preference probability can be used in the pairwise ranking methods.

13.2.1.3 Dynamic Bayesian Network Click Model

To better model the perceived relevance and actual relevance, a Dynamic Bayesian Network (DBN) model is proposed in [7]. In this model, a click is assumed to occur if and only if the user has examined the URL and deemed it relevant. Furthermore, the DBN model assumes that users make a linear transversal through the results and decide whether to click based on the perceived relevance of the document. The user chooses to examine the next URL if he/she is unsatisfied with the clicked URL (based on actual relevance). The DBN model can be represented by Fig. 13.3.

For simplicity, suppose we are only interested in the top 10 documents appearing in the first page of the search results, which means that the sequence in Fig. 13.3 goes from 1 to 10. The variables inside the box are defined at the session level, while those out of the box are defined at the query level.

Fig. 13.3 The DBN model

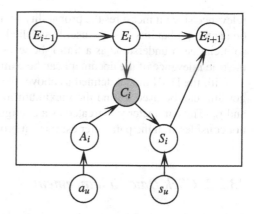

For a given position i, in addition to the observed variable C_i indicating whether there is a click or not at this position, the following hidden binary variables are defined to model examination, perceived relevance, and actual relevance, respectively:

- E_i: did the user examine the URL?
- A_i: was the user attracted by the URL?
- S_i: was the user satisfied by the actual document?

The model in Fig. 13.3 can be mathematically described as below.

$$
\begin{aligned}
A_i = 1, \qquad E_i = 1 &\Leftrightarrow C_i = 1, \\
P(A_i = 1) &= a_u, \\
P(S_i = 1 | C_i = 1) &= s_u, \\
C_i = 0 &\Rightarrow S_i = 0, \\
S_i = 1 &\Rightarrow E_{i+1} = 0, \\
P(E_{i+1} = 1 | E_i = 1, S_i = 0) &= \gamma, \\
E_i = 0 &\Rightarrow E_{i+1} = 0.
\end{aligned}
\tag{13.11}
$$

A detailed explanation of the model now follows. First one assumes that there is a click if and only if the user looks at the URL and is attracted by it. The probability of being attracted depends only on the URL. Second, the user scans the URLs linearly from top to bottom until he/she decides to stop. After the user clicks and visits the URL, there is a certain probability that he/she will be satisfied by this URL. On the other hand, if he/she does not click, he/she will not be satisfied. Once the user is satisfied by the URL he/she has visited, he/she stops his/her search. If the user is not satisfied by the current result, there is a probability $1 - \gamma$ that the user abandons his/her search and a probability γ that the user examines the next URL. If the user does not examine the position i, he/she will not examine the subsequent positions.

Note that, unlike previous models, the DBN model has two variables a_u and s_u related to the relevance of the document. The first one models the perceived relevance since it measures the probability of a click based on the URL. The second one is the probability that the user is satisfied given that he has clicked on the link; so it can been understood as a 'ratio' between actual and perceived relevance, and the true relevance of the document can be computed as $a_u s_u$.

With the DBN model defined as above, the Expectation-Maximization (EM) algorithm can be used to find the maximum likelihood estimate of the variables a_u and s_u. The parameter γ is treated as a configurable parameter for the model and is not considered in the parameter estimation process.

13.2.2   Click Data Enhancement

In the previous subsection, we have introduced various click models for ground truth mining. These models can be effective, however, they also have certain limitations. First, although the click information is very helpful, it is not the only information source that can be used to mine ground-truth labels. For example, the content information about the query and the clicked documents can also be very helpful. More reliable labels are expected to be mined if one can use more comprehensive information for the task. Second, it is almost unavoidable that the mined labels from click-through logs are highly sparse. There may be three reasons: (i) the click-through logs from a search engine company may not cover all the users' behaviors due to its limited market share; (ii) since the search results provided by existing search engines are far from perfect, it is highly possible that no document is relevant with respect to some queries and therefore there will be no clicks for such queries; (iii) users may issue new queries constantly, and therefore historical click-through logs cannot cover newly issued queries.

To tackle the aforementioned problem, in [1], Agichtein et al. consider more information to learn user interaction model using training data, and in [15], some smoothing techniques are used to expand the sparse click data. We will introduce these two pieces of work in detail in this subsection.

13.2.2.1   Learning a User Interaction Model

In [1], a rich set of features is used to characterize whether a user will be satisfied with a web search result. Once the user has submitted a query, he/she will perform many different actions (e.g., reading snippets, clicking results, navigating, and refining the query). To capture and summarize these actions, three groups of features are used: query-text, click-through, and browsing.

- *Query-text features*: Users decide which results to examine in more detail by looking at the result title, URL, and snippet. In many cases, looking at the original document is not even necessary. To model this aspect of user experience, features that characterize the nature of the query and its relation to the snippet text are extracted, including overlap between the words in the title and in the query, the fraction of words shared by the query and the snippet, etc.

- *Browsing features*: These features are used to characterize users' interactions with pages beyond the search result page. For example, one can compute how long users dwell on a page or domain. Such features allow us to model intra-query diversity of the page browsing behavior (e.g., navigational queries, on average, are likely to have shorter page dwell time than transactional or informational queries).
- *Click-through features*: Clicks are a special case of user interaction with the search engine. Click-through features used in [1] include the number of clicks for the result, whether there is a click on the result below or above the current URL, etc.

Some of the above features (e.g., click-through features and dwell time) are regarded as biased and only probabilistically related to the true relevance. Such features can be represented as a mixture of two components, one is the prior "background" distribution for the value of the feature aggregated across all queries, and the other is the component of the feature influenced by the relevance of the documents. Therefore, one can subtract the background distribution from the observed feature value for the document at a given position. This treatment can well deal with the position bias in the click-through data.

Given the above features (with the subtraction of the background distribution), a general implicit feedback interpretation strategy is learned automatically instead of relying on heuristics or insights. The general approach is to train a classifier to induce weights for the user behavior features, and consequently derive a predictive model of user preferences. The training is done by comparing a wide range of implicit behavior features with explicit human judgments for a set of queries. RankNet [4] is used as the learning machine.

According to the experiments conducted in [1], by using the machine learning based approach to combine multiple pieces of evidence, one can mine more reliable ground-truth labels for documents than purely relying on the click-through information.

13.2.2.2 Smoothing Click-Through Data

In order to tackle the sparseness problem with the click-through data, in [15], a query clustering technique is used to smooth the data.

Suppose we have obtained click-through information for query q and document d. The basic idea is to propagate the click-through information to other similar queries. In order to determine the similar queries, the co-click principle (queries for which users have clicked on the same documents can be considered to be similar) is employed. Specifically, a random walk model is used to derive the query similarity in a dynamic manner.

For this purpose, a click graph that is a bipartite-graph representation of click-through data is constructed. $\{q_i\}_{i=1}^n$ represents a set of query nodes and $\{d_j\}_{j=1}^m$ represents a set of document nodes. Then the bipartite graph can be represented by a $m \times n$ matrix W, in which $W_{i,j}$ represents the click information associated with (q_i, d_j). This matrix is then normalized to be a query-to-document transition matrix, denoted by A, where $A_{i,j}$ is the probability that q_i transits to d_j in one hop. Similarly, the transpose of matrix W can be normalized to be a document-to-query transition matrix B. Using A and B, one can compute the probability of transiting from any node to another node in the bipartite graph. Then, the query similarity is defined as the probability that one query transits to another query in two hops (or the corresponding elements in the matrix AB).

Experimental results in [15] have shown that the smooth technique is effective, and training with the smoothed click-through data can lead to the gain of ranking performance.

Note that query clustering has been well studied in the literature of Web search. Therefore, the method proposed in [15] is not the only method that one can leverage. Please also see [29, 30, 34] for more work on query clustering.

13.3 Training Data Selection

In the previous section, we have discussed general methodologies to mine ground truth labels for learning to rank from click-through data. In Chap. 1, we have also introduced the labeling methodology with human annotators. Basically both the mining and labeling methodologies assume that the data (either the click-through data or the document in the annotation pool) are available in advance, and one deals with the data (either performing mining or conducting labeling) in a passive way. However, this might not be the optimal choice because some assumptions in the mining or labeling process might not be considered when collecting the data. This will make the ground-truth labels obtained from these data less effective for training a ranking model.

Here is an example about the labeling of TREC Million Query Track. When forming the pool for judgment, those documents that can best distinguish different runs of the participants were selected. As a result, when the human assessors labeled the documents in the pool, those highly relevant documents that are ranked high by all the runs will be missing since they are not in the pool at all. Although this will not hurt evaluation (whose goal is to distinguish different methods), it will hurt the model training (whose goal is to learn as much information about relevance as possible from the training data). As a consequence, it is possible that not all the judged documents are useful for training, and some useful documents are not judged.

To tackle the aforementioned problems, it is better to think about the strategy, in addition to the methodology, for mining or labeling. That is, can we present those documents more suitable for training models but not for distinguishing methods to the human annotators? Can we pre-define the distribution of the labeled data before labeling (e.g., the number of queries versus the number of documents)? Should we use all the labeled data, or conduct some selection before training? In this section, we will introduce some previous work that tries to address the above issues.

13.3.1 Document and Query Selection for Labeling

No matter how the labels are obtained, the process is non-trivial and one needs to consider how to make it more cost-effective. There are at least two issues to be considered for this purpose. First, if we can only label a fixed total number of documents, how should we distribute them (more queries and fewer documents per query vs. fewer queries and more documents per query)? Second, if we can only label a fixed total number of documents, which of the documents in the corpus should we present to the annotators?

13.3.1.1 Deep Versus Shallow Judgments

In [32], an empirical study is conducted regarding the influence of label distribution on learning to rank. In the study, LambdaRank [11] is used as the learning-to-rank algorithm, and a dataset from a commercial search engine is used as the experimental platform. The dataset contains 382 features and is split into training, validation, and test sets with 2,000, 1,000, and 2,000 queries respectively. The average number of judged documents in the training set is 350 per query, and the number highly varies across different queries.

To test the effect of judging more queries versus more documents per query, different training sets are formed by (i) sampling p% queries while keeping the number of documents per query fixed to the maximum available, and (ii) sampling p% of documents per query and keeping the number of queries fixed. Then LambdaRank is trained using different training data and NDCG@10 on the test set is computed. The experiments are repeated ten times and the average NDCG@10 value is used for the final study.

According to the experimental results, one has the following observations.

- Given limited number of judgments, it is better to judge more queries but fewer documents per query than fewer queries with more documents per query. Sometimes additional documents per query do not result in any additional improvements in the quality of the training set.
- The lower bound on the number of documents per query is 8 on the dataset used in the study. When the lower bound is met, if one has to decrease the total number of judgments further, it is better to decrease the number of queries in the training data.

The explanation in [32] on the above experimental findings is based on the informativeness of the training set. Given some number of judged documents per query, judging more documents for this query does not really add much information to the training set. However, including a new query is much more informative since the new query may have quite different properties than the queries that are already in the training set. In [8], a theoretical explanation on this empirical finding is provided based on the statistical learning theory for ranking. Please refer to Chap. 17 for more details.

13.3.1.2 Actively Learning for Labeling

The idea of active learning is to control what is presented to the annotators (or users in the case of click-through log mining). In this subsection, we will introduce three pieces of work. The first two are designed for active learning in human annotation, and the last one is designed for active learning in click-through log mining.

In [33], the most ambiguous set of documents is selected for labeling. As we know, the support vectors in Ranking SVM [20, 21] are the document pairs that are the closest to the hyperplane of the model and thus the most ambiguous for learning. Therefore, this sampling principle can quickly identify the support vectors and reduce the total number of labeled documents to achieve a high ranking accuracy. The method is referred to as SVM selective sampling.

The basic idea of SVM selective sampling is as follows. Suppose at the ith round, we are going to select a set of documents S_i from the entire unlabeled dataset for labeling. The criterion here is that any two documents in the selected set are difficult to distinguish from each other (we denote this criterion as $C(S_i)$ for ease of reference). To find the set according to this criterion is a typical combinatorial problem and thus very difficult to solve. To tackle the challenge, a theorem is given in [33] showing that a subset of R that optimizes the selection criterion $C(S_i)$ must be consecutive. And there exists an efficient way to find the optimal set by only searching over the consecutive subsets in R. The overall complexity of the proposed new method is in the linear order of the number of unlabeled documents.

In [12], a novel criterion is proposed to select document for labeling. That is, the aim is to select a document x such that when added to the training set with a chosen label y, the model trained on the new set would have less expected error on the test set than labeling any other document. Since before labeling, no one knows the label y, an expectation is taken over all possible y as the final selection criterion.

The above criterion looks very intuitive and reasonable, however, it is not feasible since it is very difficult to really compute the expected error. Alternatively, a method is proposed to estimate how likely the addition of a new document will result in the lowest expected error on the test set without any re-training on the enlarged training set. The basic idea is to investigate the likelihood of a document to change the current hypothesis significantly. The rationality of using this likelihood lies in the following aspects: (i) adding a new data point to the labeled set can only change the error on the test set if it changes the current hypothesis; (ii) the more significant the change, the greater chance to learn the true hypothesis faster.

To realize the above idea, active sampling methods are designed for both Ranking SVM [20, 21] and RankBoost [14]. Suppose the loss change with respect to the addition of document x is $D(w, x)$, and the minimization of this loss change will result in a new model parameter w'. Then w' is compared to the ranking model learned in the previous round without document x. Basically the larger the difference is, the more likely x will be effective in changing the hypothesis. The methods have been tested on the LETOR benchmark datasets and promising results are obtained.

The limitation of the work is that one never knows whether the most significant change to the original hypothesis is good or not before the document is labeled, since changing the hypothesis to a wrong direction may even increase the expected test error.

In [27], different ways of presenting search results to users are studied in order to actively control the click-through data and get as much informative labels as possible.

According to [27], the limitations of passively collected click-through data are as follows. After submitting a query, users very rarely evaluate results beyond the first page. As a result, the click-through data that are obtained are strongly biased toward documents already ranked on the top. Highly relevant results that are not initially ranked on the top may never be observed and evaluated.

To avoid this presentation effect, the ranking presented to users should be optimized to obtain useful data, rather than strictly in terms of estimated document relevance. A naive approach is to intentionally present unevaluated results in the top few positions, aiming to collect more feedback on them. However, such an ad-hoc approach is unlikely to be useful in the long run and would hurt user satisfaction. To tackle the problem, in [27], modifications of the search results are systematically discussed, which do not substantially reduce the quality of the ranking shown to users, but produce much more informative user feedback.

In total four different modification strategies are studied and evaluated in [27]:

- *Random exploration*: Select a random pair of documents and present them first and second, then rank the remaining documents according to the original ranking results.
- *Largest expected loss pair*: Select the pair of documents d_i and d_j that have the largest pairwise expected loss contribution, and present them first and second. Then rank the remaining documents according to the original ranking results.
- *One step lookahead*: Find the pair of documents whose contribution to the expected loss is likely to decrease most after getting users' feedback, and present them the first and second.
- *Largest expected loss documents*: For each document d_i, compute the total contribution of all pairs including d_i to the expected loss of the ranking. Present the two documents with the highest total contributions at the first and second positions, and rank the remainder according to the original ranking results.

According to the experimental results in [27], as compared to the passive collection strategy used in the previous work, the last three active exploration strategies lead to more informative feedback from the users, and thus much faster convergence of the learning-to-rank process. This indicates that actively controlling what to present to the users can help improve the quality of the click-through data.

13.3.2 *Document and Query Selection for Training*

Suppose we have already obtained the training data, whose labels are either from human annotators or from click-through log mining. Now the question is as follows.

If we do not use all the data but only a proportion of it, how should we select the documents in order to maximize the effectiveness of the ranking model learned from the data? This is a meaningful question in the following sense.

- Sometimes one suffers from the scalability of the learning algorithms. When the algorithms cannot make use of the large amount of training data (e.g., out of memory), the most straightforward way is to down sample the training set.
- Sometimes the training data may contain noise or outliers. In this case, if using the entire training data, the learning process might not converge and/or the effectiveness of the learned model may be affected.

In this subsection, we will introduce some previous work that investigates the related issues. Specifically, in [2], different document selection strategies originally proposed for evaluation are studied in the context of learning to rank. In [16], the concept of pairwise preference consistency (PPC) is proposed, and the problem of document and query selection is modeled as an optimization problem that optimizes the PPC of the selected subset of the original training data.

13.3.2.1 Document Selection Strategies

In order to understand the influence of different document selection strategies on learning to rank, six document selection strategies widely used in evaluation are empirically investigated in [2]:

- *Depth-k pooling*: According to the depth pooling, the union of the top-k documents retrieved by each retrieval system submitted to TREC in response to a query is formed and only the documents in this depth-k pool are selected to form the training set.
- *InfAP sampling*: InfAP sampling [31] utilizes uniform random sampling to select documents to be judged. In this manner, the selected documents are the representatives of the documents in the complete collection.
- *StatAP sampling*: In StatAP sampling [26], with a prior of relevance induced by the evaluation measure AP, each document is selected with probability roughly proportional to its likelihood of relevance.
- *MTC*: MTC [6] is a greedy on-line algorithm that selects documents according to how informative they are in determining whether there is a performance difference between two retrieval systems.
- *Hedge*: Hedge is an on-line learning algorithm used to combine expert advices. It aims at choosing documents that are most likely to be relevant [3]. Hedge finds many relevant documents "common" to various retrieval systems.
- *LETOR*: In LETOR sampling [25], documents in the complete collection are first ranked by their BM25 scores for each query and the top-k documents are then selected.

In order to compare these sampling strategies, in the experiments of [2], for each query, documents from the complete collection are selected with different percentages from 0.6% to 60%, forming different sized subsets of the complete collection according to each strategy. Five learning-to-rank algorithms are used for evaluation: RankBoost [14], Regression [9], Ranking SVM [20, 21], RankNet [4], and LambdaRank [11]. According to the experimental results, one has the following observations:

- With some sampling strategies, training datasets whose sizes are as small as 1% to 2% of the complete collection are just as effective for learning-to-rank purposes as the complete collection. This indicates that it is not necessary to use the entire dataset for training.
- Hedge seems to be a less effective sampling strategy. Ranking functions trained on datasets constructed according to the hedge methodology only reach their optimal performance when trained over data sets that are at least 20% of the complete collection, while in the worst case, the performances of some ranking functions are significantly lower than the optimal one even when trained over 40% to 50% of the complete collection (e.g., the performances of RankBoost, Regression, and RankNet with a hidden layer).
- The other sampling strategies work fairly well,[1] though they may perform a little worse when some learning-to-rank algorithms are used. For example, the LETOR strategy does not perform very well when Ranking SVM is used, MTC does not perform very well when RankBoost and Regression are used, and infAP does not perform very well when RankBoost and RankNet with a hidden layer are used. However, overall speaking, their performances are acceptable and not so different from each other, especially when the sampling ratio is larger than 20%.

13.3.2.2 Data Selection by Optimizing PPC

In [16], Geng et al. argue that in order to improve the training performance through data selection, one needs to first define a reasonable measure of the data quality. Accordingly, a measure called *pairwise preference consistency* (PPC) is proposed, whose definition is

$$PPC(S) = \sum_{q,q'} \frac{1}{\tilde{m}_q \tilde{m}_{q'}} \sum_{u,v} \sum_{u',v'} \text{sim}\big((x_u^q - x_v^q), (x_{u'}^{q'} - x_{v'}^{q'})\big), \qquad (13.12)$$

where S is the training data collection, and sim(.) is a similarity function, a simple yet effective example is the inner product.

Then let us see how to effectively optimize the PPC of the selected subset of data. In particular, variable α_u^q is used to indicate whether document x_u^q is selected or not

---

[1]Note that the authors of [2] mis-interpreted their experimental results in the original paper. They claimed that the LETOR strategy was the second worst; however, according to the figure they showed in the paper, the LETOR strategy performed very well as compared to other strategies.

(i.e., if $\alpha_u^q = 1$, then the document is selected, and otherwise not selected). Only when both documents associated with a query are selected, can the corresponding document pair be selected. Therefore, the PPC of the selected subset can be calculated as follows (here the inner product is used as the similarity function),

$$\sum_{q,q'} \frac{1}{\tilde{m}_q \tilde{m}_{q'}} \sum_{u,v} \sum_{u',v'} \left(x_u^q - x_v^q\right)^T \left(x_{u'}^{q'} - x_{v'}^{q'}\right) \alpha_u^q \alpha_v^q \alpha_{u'}^{q'} \alpha_{v'}^{q'}. \tag{13.13}$$

The task is to find optimal indicator variables α_u^q, such that the PPC of the selected subset can be maximized. To tackle this task, an efficient stagewise solution is proposed. That is, one first finds the optimal selection of document pairs, and then take it as a given condition to find the optimal selection of documents.

Specifically, the sub-task of document pair selection is formulated as follows. Given a training data collection S, we use variable $\omega_{u,v}^q$ to indicate whether the document pair (x_u^q, x_v^q) is selected or not. Then the PPC of the selected subset of document pairs can be written as follows:

$$PPC(S, \omega) = \sum_{q,q'} \frac{1}{\tilde{m}_q \tilde{m}_{q'}} \sum_{u,v} \sum_{u',v'} \omega_{u,v}^q \omega_{u',v'}^{q'} \left(x_u^q - x_v^q\right)^T \left(x_{u'}^{q'} - x_{v'}^{q'}\right) = \omega^T R^T R \omega,$$

$$\tag{13.14}$$

where ω is a vector of dimension p ($p = \sum_q \tilde{m}_q$ is the total number of document pairs); its element is $\omega_{u,v}^q$; R is a $f \times p$ matrix (f is the number of features) with each column representing $\frac{x_u^q - x_v^q}{\tilde{m}_q}$.

Considering that reducing the size of the training data may potentially hurt the generalization ability of learning algorithms, the authors of [16] propose maximizing the number of selected document pairs at the same time as maximizing the above PPC. As a result, the following optimization problem is obtained:

$$\max_{\omega} \omega^T R^T R \omega + \mu^2 \left(e^T \omega\right)^2, \tag{13.15}$$

where the parameter μ is a tradeoff coefficient.

In order to efficiently solve the above optimization problem, the following strategies are used in [16]:

- Conducting eigendecomposition on $R^T R$.
- Projecting variable ω to the linear space spanned by the eigenvectors, so as to transform the original problem to its equivalent form.
- Solving the equivalent problem by maximizing a lower bound of its objective.

After obtaining the optimal selection of document pairs, the sub-task of document selection is formulated as below.

$$\min \sum_u \sum_v \left(\alpha_u^q \alpha_v^q - \omega_{u,v}^q\right)^2, \tag{13.16}$$

where m_q denotes the total number of documents associated with query q.

In order to efficiently solve this optimization problem, the authors of [16] relax the integer constraints on α_u and show that the relaxed problem becomes an eigenvalue decomposition problem, which can be efficiently obtained by many state-of-the-art eigensolves.

Experimental results on a commercial dataset show that the aforementioned approach to data selection works quite well, and the selected subset can lead to a ranker with significantly better ranking performance than the original training data.

13.3.3 Feature Selection for Training

Similar to document selection, the selection of features may also influence the effectiveness and efficiency of the learning-to-rank algorithms.

- A large number of features will influence the efficiency of the learning-to-rank algorithms. When this becomes an issue, the most straightforward solution is to remove some less effective features.
- Some features are not very useful and sometimes are even harmful to learning-to-rank algorithms. In this case, if using the entire feature set, the effectiveness of the learned model may be affected.

While there have been extensive studies on feature selection in classification, the study on feature selection for ranking is still limited. In [17], Geng et al. argue that it is not a good choice to directly apply the feature selection techniques for classification to ranking and propose a new feature selection method specifically designed for ranking. Basically two kinds of information are considered in the method: the importance of individual features and similarity between features.

The importance of each feature is determined using an evaluation measure (e.g., MAP and NDCG) or a loss function (e.g., loss functions in Ranking SVM [20, 21], MCRank [23], or ListNet [5]). In order to get such importance, one first ranks the documents using the feature, and then evaluates the performance in terms of the evaluation measure or the loss function. Note that for some features larger values correspond to higher ranks while for other features smaller values correspond to higher ranks. When calculating the importance, it is necessary to sort the documents twice (in the normal order and in the inverse order).

The similarity between features is used to remove redundancy in the selected features. In [17], the similarity between two features is computed on the basis of their ranking results. That is, each feature is regarded as a ranking model, and the similarity between two features is represented by the similarity between the ranking results that they produce. Many methods can be used to measure the distance between two ranking results. Specifically, Kendall's τ [22] is chosen in [17].

Considering the above two aspects, the overall feature selection criterion is formalized as the following optimization problem. That is, one selects those features with the largest total importance scores and the smallest total similarity scores.

$$\max \sum_i w_i \xi_i,$$

$$\min \sum_i \sum_{j \neq i} e_{i,j} \xi_i \xi_j, \tag{13.17}$$

$$\text{s.t.} \quad \xi_i \in \{0, 1\} \quad \text{and} \quad \sum_i \xi_i = t,$$

where t denotes the number of selected features, ξ_i indicates that the ith feature is selected, w_i denotes the importance score of the ith feature, and $e_{i,j}$ denotes the similarity between the ith and jth features.

Since the above multi-objective optimization problem is not easy to solve, it is converted to a single-objective optimization problem by linearly combining the original two objectives. Then, a greedy search method is used to optimize the new problem as follows.

- Construct an undirected graph G, in which each node represents a feature, the weight of the ith node is w_i and the weight of an edge between the ith and jth nodes is $e_{i,j}$.
- Follow the below steps in an iterative manner.
  - Select the node with the largest weight. Without loss of generality, suppose that the selected node is k_i.
  - A punishment is conducted on all the other nodes according to their similarities with the selected node. That is $w_j \leftarrow w_j - e_{k_i,j} \cdot 2c$, where c is a constant.
  - Remove node k_i from graph G together with all the edges connected to it, and put it into the selected feature set.
- Output the selected features.

According to the experiments in [17], the above method is both effective and efficient in selecting features for learning to rank.

13.4 Summary

In this chapter, we have discussed the data issue in learning to rank. Data is the root of an algorithm. Only if we have informative and reliable data can we get effective ranking models. The studies on the data issue are still limited. More research on this topic will greatly advance the state of the art of learning to rank.

13.5 Exercises

13.1 Compare different user click models introduced in this chapter, and discuss their pros and cons.

13.2 Click-through data captures informative user behaviors. Enumerate other uses of the click-through data in addition to ground-truth label mining.

13.3 Active learning has been well studied for classification. Please analyze the unique properties of ranking that make these previous techniques not fully applicable to ranking.

13.4 In this chapter, we have introduced how to select documents per query for more efficient labeling and more effective learning. Actually it is also meaningful to select queries for both labeling and learning. Design a query selection algorithm and use the LETOR benchmark datasets to test its performance.

References

1. Agichtein, E., Brill, E., Dumais, S.T., Ragno, R.: Learning user interaction models for predicting web search result preferences. In: Proceedings of the 29th Annual International ACM SIGIR Conference on Research and Development in Information Retrieval (SIGIR 2006), pp. 3–10 (2006)
2. Aslam, J.A., Kanoulas, E., Pavlu, V., Savev, S., Yilmaz, E.: Document selection methodologies for efficient and effective learning-to-rank. In: Proceedings of the 32nd Annual International ACM SIGIR Conference on Research and Development in Information Retrieval (SIGIR 2009), pp. 468–475 (2009)
3. Aslam, J.A., Pavlu, V., Savell, R.: A unified model for metasearch and the efficient evaluation of retrieval systems via the hedge algorithm. In: Proceedings of the 26th Annual International ACM SIGIR Conference on Research and Development in Information Retrieval (SIGIR 2003), pp. 393–394 (2003)
4. Burges, C.J., Shaked, T., Renshaw, E., Lazier, A., Deeds, M., Hamilton, N., Hullender, G.: Learning to rank using gradient descent. In: Proceedings of the 22nd International Conference on Machine Learning (ICML 2005), pp. 89–96 (2005)
5. Cao, Z., Qin, T., Liu, T.Y., Tsai, M.F., Li, H.: Learning to rank: from pairwise approach to listwise approach. In: Proceedings of the 24th International Conference on Machine Learning (ICML 2007), pp. 129–136 (2007)
6. Carterette, B., Allan, J., Sitaraman, R.: Minimal test collections for retrieval evaluation. In: Proceedings of the 29th Annual International ACM SIGIR Conference on Research and Development in Information Retrieval (SIGIR 2006), pp. 268–275 (2006)
7. Chapelle, O., Zhang, Y.: A dynamic Bayesian network click model for web search ranking. In: Proceedings of the 18th International Conference on World Wide Web (WWW 2009), pp. 1–10 (2009)
8. Chen, W., Liu, T.Y., Ma, Z.M.: Two-layer generalization analysis for ranking using rademacher average. In: Lafferty, J., Williams, C.K.I., Shawe-Taylor, J., Zemel, R., Culotta, A. (eds.) Advances in Neural Information Processing Systems 23 (NIPS 2010), pp. 370–378 (2011)
9. Cossock, D., Zhang, T.: Subset ranking using regression. In: Proceedings of the 19th Annual Conference on Learning Theory (COLT 2006), pp. 605–619 (2006)
10. Craswell, N., Zoeter, O., Taylor, M., Ramsey, B.: An experimental comparison of click position-bias models. In: Proceedings of the 1st International Conference on Web Search and Web Data Mining (WSDM 2008), pp. 87–94 (2008)
11. Donmez, P., Svore, K.M., Burges, C.J.C.: On the local optimality of lambdarank. In: Proceedings of the 32nd Annual International ACM SIGIR Conference on Research and Development in Information Retrieval (SIGIR 2009), pp. 460–467 (2009)

12. Donmez, R., Carbonell, J.G.: Optimizing estimated loss reduction for active sampling in rank learning. In: Proceedings of the 25th International Conference on Machine Learning (ICML 2008), pp. 248–255 (2008)
13. Dupret, G., Piwowarski, B.: A user browsing model to predict search engine click data from past observations. In: Proceedings of the 31st Annual International Conference on Research and Development in Information Retrieval (SIGIR 2008), pp. 331–338 (2008)
14. Freund, Y., Iyer, R., Schapire, R., Singer, Y.: An efficient boosting algorithm for combining preferences. Journal of Machine Learning Research **4**, 933–969 (2003)
15. Gao, J., Yuan, W., Li, X., Deng, K., Nie, J.Y.: Smoothing clickthrough data for web search ranking. In: Proceedings of the 32st Annual International Conference on Research and Development in Information Retrieval (SIGIR 2009), pp. 355–362 (2009)
16. Geng, X., Liu, T.Y., Qin, T., Cheng, X.Q., Li, H.: Selecting optimal subset for learning to rank. Information Processing and Management (2011)
17. Geng, X.B., Liu, T.Y., Qin, T., Li, H.: Feature selection for ranking. In: Proceedings of the 30th Annual International ACM SIGIR Conference on Research and Development in Information Retrieval (SIGIR 2007), pp. 407–414 (2007)
18. Guo, F., Liu, C., Kannan, A., Minka, T., Taylor, M., Wang, Y.M., Faloutsos, C.: Click chain model in web search. In: Proceedings of the 18th International Conference on World Wide Web (WWW 2009), pp. 11–20 (2009)
19. Guo, F., Liu, C., Wang, Y.M.: Efficient multiple-click models in web search. In: Proceedings of the 1st International Conference on Web Search and Web Data Mining (WSDM 2008), pp. 124–131 (2009)
20. Herbrich, R., Obermayer, K., Graepel, T.: Large margin rank boundaries for ordinal regression. In: Advances in Large Margin Classifiers, pp. 115–132 (2000)
21. Joachims, T.: Optimizing search engines using clickthrough data. In: Proceedings of the 8th ACM SIGKDD International Conference on Knowledge Discovery and Data Mining (KDD 2002), pp. 133–142 (2002)
22. Kendall, M.: Rank Correlation Methods. Oxford University Press, London (1990)
23. Li, P., Burges, C., Wu, Q.: McRank: Learning to rank using multiple classification and gradient boosting. In: Advances in Neural Information Processing Systems 20 (NIPS 2007), pp. 845–852 (2008)
24. Liu, C., Guo, F., Faloutsos, C.: Bbm: Bayesian browsing model for petabyte-scale data. In: Proceedings of the 15th ACM SIGKDD International Conference on Knowledge Discovery and Data Mining (KDD 2009), pp. 537–546 (2009)
25. Liu, T.Y., Xu, J., Qin, T., Xiong, W.Y., Li, H.: LETOR: benchmark dataset for research on learning to rank for information retrieval. In: SIGIR 2007 Workshop on Learning to Rank for Information Retrieval (LR4IR 2007) (2007)
26. Pavlu, V.: Large scale ir evaluation. Ph.D. thesis, Northeastern University, College of Computer and Information Science (2008)
27. Radlinski, F., Joachims, T.: Active exploration for learning rankings from clickthrough data. In: Proceedings of the 13th ACM SIGKDD International Conference on Knowledge Discovery and Data Mining (KDD 2007) (2007)
28. Richardson, M., Dominowska, E., Ragno, R.: Predicting clicks: estimating the click-through rate for new ads. In: Proceedings of the 16th International Conference on World Wide Web (WWW 2007), pp. 521–530 (2007)
29. Wen, J.R., Nie, J.Y., Zhang, H.J.: Query clustering using user logs. ACM Transactions on Information Systems **20**(1), 59–81 (2002)
30. Yi, J., Maghoul, F.: Query clustering using click-through graph. In: Proceedings of the 18th International Conference on World Wide Web (WWW 2009), pp. 1055–1056 (2009)
31. Yilmaz, E., Aslam, J.A.: Estimating average precision with incomplete and imperfect judgments. In: Proceedings of the Fifteenth ACM International Conference on Information and Knowledge Management (CIKM 2006), pp. 102–111 (2006)
32. Yilmaz, E., Robertson, S.: Deep versus shallow judgments in learning to rank. In: Proceedings of the 32st Annual International Conference on Research and Development in Information Retrieval (SIGIR 2009), pp. 662–663 (2009)

33. Yu, H.: Svm selective sampling for ranking with application to data retrieval. In: Proceedings of the 11th ACM SIGKDD International Conference on Knowledge Discovery and Data Mining (KDD 2005), pp. 354–363 (2005)
34. Zhao, Q., Hoi, S.C.H., Liu, T.Y., Bhowmick, S.S., Lyu, M.R., Ma, W.Y.: Time-dependent semantic similarity measure of queries using historical click-through data. In: Proceedings of the 15th International Conference on World Wide Web (WWW 2005), pp. 543–552 (2006)

Chapter 14
Applications of Learning to Rank

Abstract In this chapter, we introduce some applications of learning to rank. The major purpose is to demonstrate how to use an existing learning-to-rank algorithm to solve a real ranking problem. In particular, we will take question answering, multimedia retrieval, text summarization, online advertising, etc. as examples, for illustration. One will see from these examples that the key step is to extract effective features for the objects to be ranked by considering the unique properties of the application, and to prepare a set of training data. Then it becomes straightforward to train a ranking model from the data and use it for ranking new objects.

14.1 Overview

Up to this chapter, we have mainly used document retrieval as an example to introduce different aspects of learning to rank. As mentioned in the beginning of the book, learning-to-rank technologies have also been used in several other applications. In this chapter, we will introduce some of these applications.

Basically, in order to use learning-to-rank technologies in an application, one needs to proceed as follows. The very first step is to construct a training set. The second step is to extract effective features to represent the objects to be ranked, and the third step is to select one of the existing learning-to-rank methods, or to develop a new learning-to-rank method, to learn the ranking model from the training data. After that, this model will be used to rank unseen objects in the test phase.

In the remainder of this chapter, we will show how learning-to-rank technologies have been successfully applied in question answering [1, 14–16], multimedia retrieval [17, 18], text summarization [10], and online advertising [5, 9]. Please note that this is by no means a comprehensive list of the applications of learning to rank.

14.2 Question Answering

Question answering (QA) is an important problem in information retrieval, which differs from document retrieval. The task of question answering is to automatically answer a question posed in natural language. Due to this difference, QA is regarded as requiring more complex natural language processing (NLP) techniques than document retrieval, and natural language search engines are sometimes regarded as the next-generation search engines.

In this section, we will review the use of learning-to-rank technologies in several QA tasks, including definitional QA, quantity consensus QA, non-factoid QA, and Why QA.

14.2.1 Definitional QA

Definitional QA is a specific task in the TREC-QA track. Given the questions of "what is X" or "who is X", one extracts answers from multiple documents and combines the extracted answers into a single unified answer. QA is ideal as a means of helping people find definitions. However, it might be difficult to realize it in practice. Usually definitions extracted from different documents describe the term from different perspectives, and thus it is not easy to combine them. A more practical way of dealing with the problem is to rank the extracted definitions according to their likelihood of being good definitions, which is called definition search [16].

For this purpose, the first step is to collect definition candidates and define reasonable features as the representation of a definition. In [16], a set of heuristic rules are used to mine possible candidates. First, all the paragraphs in a document collection are extracted. Second, the $\langle term \rangle$ of each paragraph is identified. Here $\langle term \rangle$ is defined as the first base noun phrase, or the combination of two base phrases separated by 'of' or 'for' in the first sentence of the paragraph. Third, those paragraphs containing the patterns of '$\langle term \rangle$ is a/an/the ·', '$\langle term \rangle$, ·, a/an/the', or '$\langle term \rangle$ is one of ·' are selected as definition candidates.

Then, a set of features are extracted for each of these definition candidates. Specifically, the following features are used.

- $\langle term \rangle$ occurs at the beginning of a paragraph.
- $\langle term \rangle$ begins with 'the', 'a', or 'an'.
- All the words in $\langle term \rangle$ begin with uppercase letters.
- The paragraph contains predefined negative words, e.g., 'he', 'she', and 'said'.
- $\langle term \rangle$ contains pronouns.
- $\langle term \rangle$ contains 'of', 'for', 'and', 'or', ','.
- $\langle term \rangle$ re-occurs in the paragraph.
- $\langle term \rangle$ is followed by 'is a', 'is an', or 'is the'.
- Number of sentences in the paragraph.
- Number of words in the paragraph.
- Number of the adjectives in the paragraph.
- Bag of words: words frequently occurring within a window after $\langle term \rangle$.

With this feature representation, a standard Ranking SVM algorithm [7, 8] is used to learn the optimal ranking function to combine these features in order to produce a ranking for the definition candidates. The above method has been tested on both intranet data and the "Gov" dataset used by TREC. The experimental results have shown that the above method can significantly outperform several non-learning baselines, including BM25, in terms of several different measures such as error rate, R-precision, and P@k.

14.2.2 Quantity Consensus QA

Quantity search is an important special case of entity search. A quantity may be a unitless number or have an associated unit like length, mass, temperature, currency, etc. TREC-QA track 2007, 2006, and 2005 have 360, 403 and 362 factoid queries, of which as many as 125, 177, and 116 queries seek quantities. As against "spot queries" seeking unique answers like date of birth, there is uncertainty about the answer for quantity consensus queries (e.g., "driving time from Beijing to Shanghai" or "battery life of iPhone 4"). To learn a reasonable distribution over an uncertain quantity, the user may need to browse thousands of pages returned by a regular search engine. This is clearly not only time consuming but also infeasible for users.

In [1], learning-to-rank technologies are applied to solve the problem of quantity consensus QA. This is, however, not an easy task. The difficulty lies in that the answer should be derived from all the documents relevant to the query through deep mining, and simple ranking of these documents cannot provide such information. To tackle this challenge, evidence in favor of candidate quantities and quantity intervals from documents (or snippets) is aggregated in a collective fashion, and these intervals instead of original documents (or snippet) are ranked.

For this purpose, one first finds candidate intervals related to the quantity consensus query. This is done by first processing each snippet and finding the quantity (including unit) that it contains. Then these quantities are put into the x-axis, and their merits are evaluated using a merit function. The function basically considers the number of quantities falling into an interval, and the relevance of the corresponding snippets to the query. The intervals with top merits are selected as candidate intervals and passed onto the learning-to-rank algorithm. Note that each selected interval is associated with several snippets whose quantities fall into this interval.

Second, one needs to extract a set of features as the representation of an candidate interval. Specifically in [1], the following features are extracted:

- Whether all snippets associated with the interval contain some query word.
- Whether all snippets associated with the interval contain the minimum IDF query word.
- Whether all snippets associated with the interval contain the maximum IDF query word.
- Number of distinct words found in snippets associated with the interval.
- Number of words that occur in all snippets associated with the interval.
- One minus the number of distinct quantities mentioned in snippets associated with the interval, divided by the length of the interval.
- Number of snippets associated with the interval, divided by the total number of snippets retrieved for the query.

• Features corresponding to the merit of the interval.

Third, with the above feature representation, standard Ranking SVM [7, 8] is used to combine these features and produce a ranked list of all the candidate intervals.

The above method has been tested on TREC-QA data, and the experimental results show significant improvement over several non-learning baselines in terms of both NDCG and MAP. Indirect comparison with the TREC participants also suggests that the proposed method is very competitive: ranked the second-best in TREC-QA 2007 data, and ranked five out of 63 teams on TREC-QA 2004 data.

14.2.3 Non-factoid QA

In [14], an answer ranking engine for non-factoid questions built using a large online community-generated question-answer collection (Yahoo! Answers) is proposed. Through the construction of the engine, two issues are investigated. (i) Is it possible to learn an answer ranking model for complex questions from noisy data? (ii) Which features are most useful in this scenario?

In the proposed engine, there are three key components, the *answer retrieval component* uses unsupervised information retrieval models, the *answer ranking component* uses learning-to-rank technologies, and the *question-to-answer translation model* uses class-conditional learning techniques. Here we mainly introduce the learning-to-rank technology used in the answer ranking component.

First of all, given a set of answers, one needs to extract features to represent their relevance to the question. In [14], the following features are extracted.

• *Similarity features:* BM25 and TF-IDF on five different representations of questions and answers: words, dependencies, generalized dependencies, bigrams, and generalized bigrams.
• *Translation features:* the probability that the question Q is a translation of the answer A, computed using IBM's model 1 [2], also on five different representations.
• *Density and frequency features:* same word sequence, answer span, overall matches, and informativeness.
• *Web correlation features:* web correlation and query log correlation.

With the above features, the learning-to-rank method proposed in [13], called Ranking Perceptron, is employed to learn the answer ranking function. The basic idea of Ranking Perceptron is as follows. Given a weight vector w the score for a candidate answer x is simply the inner product between x and w, i.e., $f(x) = w^T x$. In training, for each pair (x_u, x_v), the score $f(x_u - x_v)$ is computed. Given a margin function $g(u, v)$ and a positive rate τ, if $f(x_u - x_v) \leq g(u, v)\tau$, an update is performed:

$$w^{t+1} = w^t + (x_u - x_v)g(u, v)\tau, \tag{14.1}$$

where $g(u, v) = (\frac{1}{u} - \frac{1}{v})$ and τ is found empirically using a validation set.

For regularization purposes, the average of all Perceptron models obtained during training is used as the final ranking model. The model has been tested on the Yahoo! QA data, in terms of P@1 and MRR. The experimental results show that the learning-to-rank method can significantly outperform several non-learning baseline methods, and better ranking performances can be achieved when more features are used in the learning-to-rank process.

14.2.4 Why QA

Why-questions are widely asked in real world. Answers to why-questions tend to be at least one sentence and at most one paragraph in length. Therefore, passage retrieval appears to be a suitable approach to Why QA.

In [15], different learning-to-rank algorithms are empirically investigated to perform the task of answering why-questions. For this purpose, the Wikipedia INEX corpus is used, which consists of 659,388 articles extracted from the online Wikipedia in the summer of 2006, converted to XML format. By applying some segmentation methods, 6,365,890 passages are generated, which are the objects to be ranked with respect to given why-questions.

For each paragraph, 37 features are extracted, including TF-IDF, 14 syntactic features describing the overlap between QA constituents (e.g., subject, verb, question focus), 14 WordNet expansion features describing the overlap between the WordNet synsets of QA constituents, one cue word feature describing the overlap between candidate answer and a predefined set of explanatory cue words, six document structure features describing the overlap between question words and document title and section heading, and one WordNet Relatedness feature describing the relatedness between questions and answers according to the WordNet similarity tool.

Based on the aforementioned data and feature representations, a number of learning-to-rank methods are examined, including the pointwise approach (Naive Bayes, Support Vector Classification, Support Vector Regression, Logistic Regression), pairwise approach (Pairwise Naive Bayes, Pairwise Support Vector Classification, Pairwise Support Vector Regression, Pairwise Logistic Regression, Ranking SVM), and listwise approach (RankGP). MRR and Success at Position 10 are used as evaluation measures.

Three factors are considered in the empirical investigation: (1) the distinction between the pointwise approach, the pairwise approach, and the listwise approach; (2) the distinction between techniques based on classification and techniques based on regression; and (3) the distinction between techniques with and without hyper parameters that must be tuned.

With respect to (1), the experimental results indicate that one is able to obtain good results with both the pointwise and the pairwise approaches. The optimum score is reached by Support Vector Regression for the pairwise representation, but some of the pointwise settings reach scores that are not significantly lower than this optimum. The explanation is that the relevance labeling of the data is on a binary scale, which makes classification feasible. The good results obtained with the listwise approach, implemented as a Genetic Algorithm that optimizes MRR, are probably due to the fact that this approach allows for optimizing the evaluation criterion directly.

With respect to (2), the experimental results indicate that the classification and regression techniques are equally capable of learning to classify the data in a point-wise setting but only if the data are balanced (by oversampling or applying a cost factor) before presenting them to a classifier. For regression techniques, balancing is not necessary and even has a negative effect on the results. The results also show that it is a good option to transform the problem to a pairwise classification task for curing the class imbalance.

With respect to (3), the experimental results indicate that for the imbalanced dataset, techniques with hyper parameters heavily depend on tuning in order to find sensible hyper parameter values. However, if the class imbalance is solved by balancing the data or presenting the problem as a pairwise classification task, then the default hyper parameter values are well applicable to the data and tuning is less important.

14.3 Multimedia Retrieval

Multimedia retrieval is an important application, which has been supported by major commercial search engines. Due to the semantic gap in multimedia data, in order to provide satisfactory retrieval results, textual information such as anchor text and surrounding text is still important in determining the ranking of the multimedia objects. To further improve the ranking performance, researchers have started to pay attention to using visual information to re-rank the multimedia objects. The goal is to maintain the text-based search paradigm while improving the search results. Basically the initial text-based search results are regarded as the pseudo-ground truth of the target semantic, and the re-ranking methods are used to mine the contextual patterns directly from the initial search result and further refine it. Several learning based re-ranking methods have been proposed in the literature of multimedia retrieval, which are basically based on pseudo-relevance feedback and classification. Most recently, learning-to-rank algorithms such as Ranking SVM [7, 8] and ListNet [4] have also been applied to the task [17, 18]. When evaluated on the TRECVID 2005 video search benchmark, these learning-to-rank methods perform much better than previous methods in the literature of multimedia retrieval.

While it is promising to apply learning to rank to multimedia retrieval, there are also some differences between standard learning to rank and multimedia re-ranking that need our attention. First, while standard learning to rank requires a great amount of supervision, re-ranking takes an unsupervised fashion and approximates the initial results as the pseudo ground truth. Second, for standard learning to rank, the ranking function is trained in advance to predict the relevance scores for arbitrary queries, while for re-ranking the function is particularly trained at runtime to compute the re-ranked relevance scores for each query itself. By considering these aspects, a re-ranking framework as described below is proposed in [17, 18].

- Visual feature selection: select informative visual features using some feature selection methods to reduce the dimensionality of the feature space.
- Employment of learning-to-rank algorithms: randomly partition the dataset into n folds. Hold one fold as the test set and train the ranking function using a learning-to-rank method on the remaining data. Predict the relevance scores of the test set. Repeat until each fold is held out for testing once. The predicted scores of different folds are combined to generate a new visual ranking score.
- Rank aggregation: After normalization, the ranking score produced by the original text retrieval function and the visual ranking score are linearly combined to produce a merged score.
- Re-Rank: Sort the combined scores to output a new ranked list for the multimedia data.

The above techniques have been tested in different scenarios of multimedia retrieval, such as image tag recommendation and multiple canonical image selection. The corresponding experiments indicate that the learning-to-rank methods can significantly boost the ranking performance over the initial text-based retrieval results, and can also outperform many heuristic based re-ranking methods proposed in the literature.

14.4 Text Summarization

It has become standard for search engines to augment result lists with document summaries. Each document summary may consist of a title, abstract, and a URL. It is important to produce high quality summaries, since the summaries can bias the perceived relevance of a document.

Document summarization can either be query independent or query dependent. A query independent summary conveys general information about the document, and typically includes a title, static abstract, and URL, if applicable. The main problem with query independent summarization is that the summary for a document never changes across queries. In contrast, query-dependent summarization biases the summary towards the query. These summaries typically consist of a title, dynamic abstract, and URL. Since these summaries are dynamically generated, they are typically constructed at query time.

In [10], the authors study the use of learning-to-rank technologies for generating a query-dependent document summary. In particular, the focus is placed on the task of selecting relevant sentences for inclusion in the summary.

For this purpose, one first needs to extract features as the representation of each sentence. In [10], the following features are extracted.

- Query-Dependent Features, including exact match, the fraction of query terms that occur in the sentence, the fraction of synonyms of query terms that occur in the sentence, and the output of the language model.
- Query Independent Features, including the total number of terms in the sentence, and the relative location of the sentence within the document.

With the above feature representation, three learning-to-rank methods are examined: Ranking SVM [7, 8], support vector regression, and gradient boosted decision tree. The first one is a pairwise ranking algorithm, while the next two are pointwise ranking methods. The TREC 2002, 2003, and 2004 novelty track data are used as the experimental datasets. R-Precision is used as the evaluation measure. The experimental results basically show that support vector regression and gradient boosted decision tree significantly outperform a simple language modeling baseline and Ranking SVM. Furthermore, gradient boosted decision tree is very robust and achieves strong effectiveness across three datasets of varying characteristics.

The above experimental findings are different from the experimental findings on other tasks such as document retrieval and question answering. This is possibly because sentence selection for summarization is by nature more like a classification problem for which the pointwise approach can work very effectively, and even more effectively than the pairwise approach.

14.5 Online Advertising

The practice of sponsored search, where the paid advertisements appear alongside web search results, is now one of the largest sources of revenue for search engine companies.

When a user types a query, the search engine delivers a list of ads that are relevant to the query adjacent to or above the search results. When a user clicks on an ad, he/she is taken to the landing page of the advertisement. Such click generates a fixed amount of revenue to search engine. Thus, the total revenue generated by a particular ad to the search engine is the number of clicks multiplied by the cost per click.

It is clear that the search engine will receive more clicks if better matched ads are delivered. Thus, a high accuracy of delivering the most preferred ads to each user will help the search engine to maximize the number of clicks. However, this is not all that the search engines want. The ultimate goal of a search engine is to find an optimal ranking scheme which can maximize the total revenue. Specifically, besides the number of clicks, an extremely important factor, which substantially influences the total revenue, is the bid price of each ad. Therefore, the best choice for defining the ad ranking function is to consider both the likelihood of being clicked and the bid price of an ad (i.e., the maximization of the revenue). There are several works that apply learning-to-rank technologies to solve the problem of revenue maximization, such as [5, 9, 19]. Here we will take [19] as an example for detailed illustration.

Once again, the first step is to extract features to represent an ad. Specifically, in [19], the following features are extracted.

- Relevance features, including term frequency, TF-IDF, edit distance, BM25 of title, BM25 of description, LMIR of title, and LMIR of description.
- Click-through features, including ad click-through rate (CTR), campaign CTR, and account CTR.
- Other features, including bid price and match type.

With the above feature representations, several learning-to-rank technologies are applied to optimize the total revenue. One of the models is call RankLogistic. The model is very similar to RankNet [3]: its loss function is also defined as a pairwise cross entropy. The difference lies in that RankLogisitic also involves a regularization item in its loss function. Another model is called revenue direct optimization. In this model, the revenue is first formulated as follows:

$$Rev(w) = \sum_{i=1}^{n} \sum_{j=1}^{m^{(i)}} r_j^{(i)} c_j^{(i)} I_{\{\min_{k \neq j} f(w, x_j^{(i)}) - f(w, x_k^{(i)}) > 0\}}, \qquad (14.2)$$

where $r_j^{(i)}$ is the bid price for the jth ad associated with query q_i, $c_j^{(i)}$ indicates whether the jth ad associated with query q_i is clicked by the user, and $I_{\{\min_{k \neq j} f(w, x_j^{(i)}) - f(w, x_k^{(i)}) > 0\}}$ measures whether the ranking function f can rank the ad on the top position.

Since the revenue defined above is not continuous, $I_{\{\min_{k \neq j} f(w, x_j^{(i)}) - f(w, x_k^{(i)}) > 0\}}$ is first replaced with $\sum_{k \neq j} I_{\{f(w, x_j^{(i)}) - f(w, x_k^{(i)}) > 0\}}$, and then approximated by changing the indicator function for a Bernoulli log-likelihood function. In this way, a continuous loss function is obtained as follows:

$$L(w) = \lambda \|w\|^2 + \sum_{i=1}^{n} \sum_{j=1}^{m^{(i)}} r_j^{(i)} c_j^{(i)} \sum_{k \neq j} \log\left(1 + e^{f(w, x_k^{(i)}) - f(w, x_j^{(i)})}\right). \qquad (14.3)$$

The above models have been tested on commercial data, with Ad CTR as the evaluation measure. The experimental results show that the revenue direct optimization method can outperform RankLogisic and several other baselines. This empirical observation is consistent with what we have observed in document retrieval: direct optimization of evaluation measures can usually outperform pairwise ranking methods.

14.6 Summary

In addition to the above applications, learning-to-rank technologies have also been applied in several other applications, such as collaborative filtering [11], expert finding [6], and subgroup ranking in database [12]. From these examples, we can see that ranking is the key problem in many different applications, and learning-to-rank technologies can help improve conventional ranking heuristics by learning the optimal ranking model from training data. This demonstrates well the practical impact of the various research efforts on learning-to-rank. We hope that learning-to-rank technologies can be applied in more and more applications, and continuously contribute to the development of information retrieval, natural language processing, and many other fields.

14.7 Exercises

14.1 Enumerate other potential applications of learning to rank in the field of computer science.

14.2 Specify a field where learning-to-rank technologies can be potentially used, and go through the entire process of applying learning-to-rank technologies to advance its state of the art. Work items include a mathematical formulation of the problem, feature extraction, loss function design, optimization method implementation, etc.

14.3 Feature extraction is a key step in using learning-to-rank technologies in an application, since it is usually the features that capture the unique characteristics of the application. Use the examples given in the chapter to discuss how the features extracted in each work reflect the characteristics of their target application.

References

1. Banerjee, S., Chakrabarti, S., Ramakrishnan, G.: Learning to rank for quantity consensus queries. In: Proceedings of the 32nd Annual International ACM SIGIR Conference on Research and Development in Information Retrieval (SIGIR 2009), pp. 243–250 (2009)
2. Brown, P., Pietra, S.D., Pietra, V.D., Mercer, R.: The mathematics of statistical machine translation: parameter estimation. Internet Mathematics 19(2), 263–311 (1993)
3. Burges, C.J., Shaked, T., Renshaw, E., Lazier, A., Deeds, M., Hamilton, N., Hullender, G.: Learning to rank using gradient descent. In: Proceedings of the 22nd International Conference on Machine Learning (ICML 2005), pp. 89–96 (2005)
4. Cao, Z., Qin, T., Liu, T.Y., Tsai, M.F., Li, H.: Learning to rank: from pairwise approach to listwise approach. In: Proceedings of the 24th International Conference on Machine Learning (ICML 2007), pp. 129–136 (2007)
5. Ciaramita, M., Murdock, V., Plachouras, V.: Online learning from click data for sponsored search. In: Proceeding of the 17th International Conference on World Wide Web (WWW 2008), pp. 227–236 (2008)
6. Fang, Y., Si, L., Mathur, A.P.: Ranking experts with discriminative probabilistic models. In: SIGIR 2009 Workshop on Learning to Rank for Information Retrieval (LR4IR 2009) (2009)
7. Herbrich, R., Obermayer, K., Graepel, T.: Large margin rank boundaries for ordinal regression. In: Advances in Large Margin Classifiers, pp. 115–132 (2000)
8. Joachims, T.: Optimizing search engines using clickthrough data. In: Proceedings of the 8th ACM SIGKDD International Conference on Knowledge Discovery and Data Mining (KDD 2002), pp. 133–142 (2002)
9. Mao, J.: Machine learning in online advertising. In: Proceedings of the 11th International Conference on Enterprise Information Systems (ICEIS 2009), p. 21 (2009)
10. Metzler, D.A., Kanungo, T.: Machine learned sentence selection strategies for query-biased summarization. In: SIGIR 2008 Workshop on Learning to Rank for Information Retrieval (LR4IR 2008) (2008)
11. Pessiot, J.F., Truong, T.V., Usunier, N., Amini, M.R., Gallinari, P.: Learning to rank for collaborative filtering. In: Proceedings of the 9th International Conference on Enterprise Information Systems (ICEIS 2007), pp. 145–151 (2007)
12. Rueping, S.: Ranking interesting subgroups. In: Proceedings of the 26th International Conference on Machine Learning (ICML 2009), pp. 913–920 (2009)
13. Shen, L., Joshi, A.K.: Ranking and reranking with perceptron. Journal of Machine Learning 60(1–3), 73–96 (2005)

14. Surdeanu, M., Ciaramita, M., Zaragoza, H.: Learning to rank answers on large online qa collections. In: Proceedings of the 46th Annual Meeting of the Association for Computational Linguistics: Human Language Technologies (ACL-HLT 2008), pp. 719–727 (2008)
15. Verberne, S., Halteren, H.V., Theijssen, D., Raaijmakers, S., Boves, L.: Learning to rank qa data. In: SIGIR 2009 Workshop on Learning to Rank for Information Retrieval (LR4IR 2009) (2009)
16. Xu, J., Cao, Y., Li, H., Zhao, M.: Ranking definitions with supervised learning methods. In: Proceedings of the 14th International Conference on World Wide Web (WWW 2005), pp. 811–819. ACM Press, New York (2005)
17. Yang, Y.H., Hsu, W.H.: Video search reranking via online ordinal reranking. In: Proceedings of IEEE 2008 International Conference on Multimedia and Expo (ICME 2008), pp. 285–288 (2008)
18. Yang, Y.H., Wu, P.T., Lee, C.W., Lin, K.H., Hsu, W.H., Chen, H.H.: Contextseer: context search and recommendation at query time for shared consumer photos. In: Proceedings of the 16th International Conference on Multimedia (MM 2008), pp. 199–208 (2008)
19. Zhu, Y., Wang, G., Yang, J., Wang, D., Yan, J., Hu, J., Chen, Z.: Optimizing search engine revenue in sponsored search. In: Proceedings of the 32nd Annual International ACM SIGIR Conference on Research and Development in Information Retrieval (SIGIR 2009), pp. 588–595 (2009)

Part VI
Theories in Learning to Rank

In this part, we will focus on the theoretical issues regarding learning to rank. Since ranking is different from regression and classification, previous learning theories cannot be directly applied to ranking and the development of new theories is needed. In this part, we will introduce the recent advances in statistical learning theory for ranking, including statistical ranking framework, generalization ability, and statistical consistency of ranking methods.

After reading this part, the readers are expected to understand the big picture of statistical learning theory for ranking, and know what has been done and what is still missing in this area.

Chapter 15
Statistical Learning Theory for Ranking

Abstract In this chapter, we introduce the statistical learning theory for ranking. In order to better understand existing learning-to-rank algorithms, and to design better algorithms, it is very helpful to deeply understand their theoretical properties. In this chapter, we give the big picture of theoretical analysis for ranking, and point out several important issues to be investigated: statistical ranking framework, generalization ability, and statistical consistency for ranking methods.

15.1 Overview

As a new machine learning problem, ranking is not only about effective algorithms but also about the theory behind these algorithms.

Actually, theoretical analysis on an algorithm always plays an important role in machine learning. This is because in practice one can only observe experimental results on relatively small-scale datasets (e.g., the experimental results on the LETOR benchmark datasets as introduced in Chap. 11). To some extent, such empirical results might not be fully reliable, because a small training set sometimes cannot fully realize the potential of a learning algorithm, and a small test set sometimes cannot reflect the true performance of an algorithm. This is due to the fact that the input and output spaces are too large to be well represented by a small number of samples. In this regard, theories are sorely needed in order to analyze the performance of a learning algorithm in the case that the training data are infinite and the test data are randomly sampled from the input and output spaces.

15.2 Statistical Learning Theory

In Fig. 15.1, we give a typical taxonomy of statistical learning theory for empirical risk minimization, which summarizes most of the theoretical aspects that one is interested in.

Basically, to understand a learning problem theoretically, we need to have a statistical framework, which describes the assumptions on the data generation (e.g., whether the instances are i.i.d.). We also need to define the true loss of the learning

T.-Y. Liu, *Learning to Rank for Information Retrieval*,
DOI 10.1007/978-3-642-14267-3_15, © Springer-Verlag Berlin Heidelberg 2011

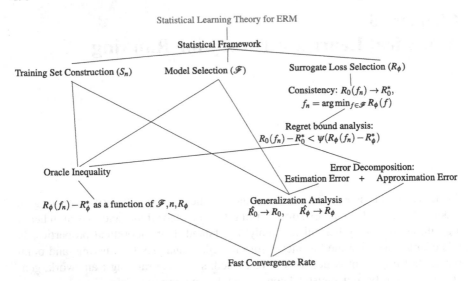

Fig. 15.1 Taxonomy of statistical learning theory

problem, which serves as a reference to study the properties of different surrogate loss functions used by various learning algorithms. Usually the average surrogate loss on the training data is called *the empirical surrogate risk* (denoted as $\hat{R}_\phi$), the expected surrogate loss on the entire product space of input and output is called *the expected surrogate risk* (denoted as R_ϕ), the average true loss on the training data is called *the empirical true risk* (denoted as $\hat{R}_0$), and the expected true loss on the entire product space of input and output is called *the expected true risk* (denoted as R_0).

With the above settings, there are three major tasks in machine learning: the selection of training data (sampled from input and output spaces), the selection of hypothesis space, and the selection of the surrogate loss function. In the figure, we refer to them as training set construction, model selection, and surrogate loss selection, respectively.

In order to select appropriate training data (e.g., the number of training instances needed to guarantee the given test performance of the learned model), one needs to study the *Generalization Ability* of the algorithm, as a function of the number of training instances. The generalization analysis on an algorithm is concerned with whether and at what rate its empirical risk will converge to the expected risk, when the number of training samples approaches infinity. Sometimes, we alternatively represent the generalization ability of an algorithm using the bound of the difference between its expected risk and empirical risk, and see whether and at what rate the bound will converge to zero when the number of training samples approaches infinity. In general, an algorithm is regarded as better than the other algorithm if its empirical risk can converge to the expected risk but that of the other cannot. Furthermore, an algorithm is regarded as better than the other if its corresponding convergence rate is faster than that of the other.

In order to select the appropriate model, one needs to investigate the complexity of the corresponding function class. If the function class is too complex, then the model from this class will likely over fit even if trained with a large number of instances. Usually a well-defined complexity will determine the uniform generalization bound for a learning problem. That is, no matter which algorithm one uses, it is always a bound of the generalization ability for the algorithm.

In order to select the appropriate surrogate loss function, one needs to study whether the optimal model learned by minimizing the empirical surrogate risk will be the same as the ranking model with the minimum expected true risk. This property is called statistical consistency. To perform consistency analysis, one usually needs to derive a regret bound to bridge the surrogate loss and the true loss, and then once again apply the result of generalization analysis to bridge the empirical risk and expected risk.

From the above discussions, one can see that the statistical framework, generalization analysis, and statistical consistency are the key issues to be addressed for a learning theory. When talking about ranking, the corresponding issues will become the statistical *ranking* framework, generalization analysis for *ranking*, and statistical consistency of *ranking* methods.

Note that the above issues are critical for all machine learning problems. However, due to the differences in different machine learning problems, the corresponding conclusions and proof techniques regarding these issues may be significantly different. Just because of this, one cannot simply extend the existing results for classification to that for ranking, and new theories and proof techniques need to be developed. In the following section, we will make brief discussions on this, and show recent advances in statistical learning theory for ranking.

15.3 Learning Theory for Ranking

In this section, we will introduce the key issues regarding statistical learning theory for ranking.

15.3.1 Statistical Ranking Framework

In the literature of learning to rank, several different statistical frameworks have been used. For example, when conducting theoretical analysis on ranking methods, some researchers assume that all the documents are i.i.d. no matter they are associated with the same query or not [1–3, 5, 7, 11]. Some researchers regard documents as deterministically given, and only treat queries as i.i.d. random variables [6, 8]. Some other researchers assume that the queries are i.i.d. in the query space, and the documents associated with the same query are conditionally i.i.d. (the distribution depends on the query) [4, 10]. These different assumptions result in different settings of learning to rank, which we call document ranking, subset ranking, and two-layer ranking, respectively. We will make detailed introductions to these three settings in the next chapter.

15.3.2 Generalization Analysis for Ranking

Given the existence of queries and documents in learning to ranking for information retrieval, when talking about the generalization ability of an algorithm, one needs to pay attention to both the number of training queries and that of training documents. For example, a natural question to answer is as follows: what are the minimum numbers of training queries and training documents in order to guarantee a target difference between training error and test error? If the total number of documents to label is constrained, shall we label more queries (shallow labeling) or more documents per query (deep labeling)?

In the literature of learning to rank, some empirical studies have been conducted to answer the above questions, however, most existing work on generalization analysis for ranking cannot answer these questions yet. As one will see in the next chapter, in these works, either generalization bounds with respect to the number of documents (or document pairs) [1–3, 5, 7, 11], or those with respect to the number of queries have been derived [8, 10]. Only in a recent work [4], an answer is given to the trade off between the number of queries and the number of documents.

15.3.3 Statistical Consistency for Ranking

As aforementioned, the statistical consistency is with respect to a true loss for ranking. Since the ranking task is more complex than classification, its true loss should also consider more factors. Partly because of this, there is no consensus on the true loss for ranking yet. Instead, different true losses are assumed in different researches, including the pointwise 0–1 loss, pairwise 0–1 loss [5, 7], permutation-level 0–1 loss [13], top-k loss [12], and measure-based ranking errors (e.g., $(1-\text{NDCG})$ and $(1-\text{MAP})$) [6]. The difficulties of theoretical analyses regarding different true losses vary largely. Those with respect to the pointwise or pairwise 0–1 loss are relatively easy because of the analogy to classification. The analyses with respect to measure-based ranking errors are the most difficult. Because such true losses are defined at the query level and are position based, their mathematical properties are not very good.

Given a true loss, in order to perform meaningful analysis on statistical consistency, one needs to derive a regret bound. When the pointwise or pairwise 0–1 loss is used as the true loss for ranking, the task is similar to that for classification. However, when more complex true loss, such as the permutation-level 0–1 loss, top-k loss, and measure-based ranking errors are used, the task will become much more complex. In such cases, one usually needs to make some assumptions on the input and output spaces, and make detailed case studies. We will show some such examples [5, 6, 9, 12, 13] in Chap. 18. Along with the difficulty, actually the analysis for ranking is also more interesting and content-rich than for classification. For example, when the top-k loss is used as the true loss [12], one can further analyze how the statistical consistency changes with respect to different k values. This is quite unique, and has never appeared in the literature of classification.

In the next three chapters, we will give detailed introductions to existing works on statistical ranking framework, generalization analysis for ranking, and statistical consistency for ranking. Note that it is unavoidable that a lot of mathematics will be used in these chapters, in order to make the discussions rigorous and clear. It is safe, however, to skip this whole part, if one only wants to know the algorithmic development of learning to rank.

15.4 Exercises

15.1 List the differences between ranking and classification, which pose potential challenges to the theoretical analysis on ranking.
15.2 Enumerate the major problems to investigate in statistical learning theory, and their relationships.

References

1. Agarwal, S.: Generalization bounds for some ordinal regression algorithms. In: Proceedings of the 19th International Conference on Algorithmic Learning Theory (ALT 2008), pp. 7–21 (2008)
2. Agarwal, S., Graepel, T., Herbrich, R., Har-Peled, S., Roth, D.: Generalization bounds for the area under the roc curve. Journal of Machine Learning 6, 393–425 (2005)
3. Agarwal, S., Niyogi, P.: Stability and generalization of bipartite ranking algorithms. In: Proceedings of the 18th Annual Conference on Learning Theory (COLT 2005), pp. 32–47 (2005)
4. Chen, W., Liu, T.Y., Ma, Z.M.: Two-layer generalization analysis for ranking using rademacher average. In: Lafferty, J., Williams, C.K.I., Shawe-Taylor, J., Zemel, R., Culotta, A. (eds.) Advances in Neural Information Processing Systems 23 (NIPS 2010), pp. 370–378 (2011)
5. Clemencon, S., Lugosi, G., Vayatis, N.: Ranking and empirical minimization of u-statistics. The Annals of Statistics 36(2), 844–874 (2008)
6. Cossock, D., Zhang, T.: Subset ranking using regression. In: Proceedings of the 19th Annual Conference on Learning Theory (COLT 2006), pp. 605–619 (2006)
7. Freund, Y., Iyer, R., Schapire, R., Singer, Y.: An efficient boosting algorithm for combining preferences. Journal of Machine Learning Research 4, 933–969 (2003)
8. Lan, Y., Liu, T.Y.: Generalization analysis of listwise learning-to-rank algorithms. In: Proceedings of the 26th International Conference on Machine Learning (ICML 2009), pp. 577–584 (2009)
9. Lan, Y., Liu, T.Y., Ma, Z.M., Li, H.: Statistical consistency of ranking methods. Tech. rep., Microsoft Research (2010)
10. Lan, Y., Liu, T.Y., Qin, T., Ma, Z., Li, H.: Query-level stability and generalization in learning to rank. In: Proceedings of the 25th International Conference on Machine Learning (ICML 2008), pp. 512–519 (2008)

11. Rajaram, S., Agarwal, S.: Generalization bounds for k-partite ranking. In: NIPS 2005 Workshop on Learning to Rank (2005)
12. Xia, F., Liu, T.Y., Li, H.: Statistical consistency of top-k ranking. In: Advances in Neural Information Processing Systems 22 (NIPS 2009), pp. 2098–2106 (2010)
13. Xia, F., Liu, T.Y., Wang, J., Zhang, W., Li, H.: Listwise approach to learning to rank—theorem and algorithm. In: Proceedings of the 25th International Conference on Machine Learning (ICML 2008), pp. 1192–1199 (2008)

Chapter 16
Statistical Ranking Framework

Abstract In this chapter, we introduce the statistical ranking framework. In order to analyze the theoretical properties of learning-to-rank methods, the very first step is to establish the right probabilistic context for the analysis. This is just what the statistical ranking framework addresses. In this chapter we will show three ranking frameworks used in the literature of learning to rank, i.e., the document ranking framework, the subset ranking framework, and the two-layer ranking framework. The discussions in this chapter set the stage for further discussions on generalization ability and statistical consistency in the following chapters.

As mentioned in the previous chapter, to facilitate the discussions on the generalization ability and statistical consistency of learning-to-rank algorithms, a statistical ranking framework is needed. The framework basically describes how the data samples are generated, and how the empirical and expected risks are defined.

In the literature of learning to rank, three different statistical ranking frameworks have been used, which we call the document ranking, subset ranking, and two-layer ranking frameworks, respectively. Even for the same data, these frameworks try to give different probabilistic interpretations of its generation process. For example, the document ranking framework regards all the documents (no matter whether they are associated with the same query or not) as i.i.d. sampled from a document space; the subset ranking framework instead assumes the queries are i.i.d. sampled from the query space, and each query is associated with a deterministic set of documents; the two-layer ranking framework assumes i.i.d. sampling for both queries and documents associated with the same query. Also, these frameworks define risks in different manners. For example, the document ranking framework defines expected risks by taking integration overall all documents; the subset ranking framework defines expected risks by taking integration overall all queries, and the two-layer ranking framework defines expected risks by taking integration overall both queries and documents.

T.-Y. Liu, *Learning to Rank for Information Retrieval*,
DOI 10.1007/978-3-642-14267-3_16, © Springer-Verlag Berlin Heidelberg 2011

16.1 Document Ranking Framework

In the document ranking framework [1–3, 5, 7, 10], only documents (together with their labels) are considered as i.i.d. instances, and there is no similar consideration on queries. With such a kind of assumption, for different approaches to learning to rank, different kinds of risks can be defined.

16.1.1 The Pointwise Approach

For the pointwise approach, suppose that (x_j, y_j) are i.i.d. random variables according to distribution P, where $x \in \mathcal{X}$ stands for the document and $y \in \mathcal{Y}$ stands for the ground-truth label of the document. Given the scoring function f, a loss occurs if the prediction given by f is not in accordance with the given label. Here we use $L(f; x_j, y_j)$ as a general representation of loss functions. It can be the pointwise 0-1 loss, or the surrogate loss functions used by various pointwise ranking algorithms. Given the loss function, the *expected risk* is defined as

$$R(f) = \int_{\mathcal{X} \times \mathcal{Y}} L(f; x_j, y_j) \, P(dx_j, dy_j). \tag{16.1}$$

Intuitively, the expected risk means the loss that a ranking model f would make for a random document. Since it is almost impossible to compute the expected risk, in practice, the empirical risk on the training set is used as an estimate of the expected risk. In particular, given the training data $\{(x_j, y_j)\}_{j=1}^{m}$, the *empirical risk* can be defined as follows:

$$\hat{R}(f) = \frac{1}{m} \sum_{j=1}^{m} L(f; x_j, y_j). \tag{16.2}$$

16.1.2 The Pairwise Approach

In the pairwise approach, document pairs are learning instances. There are two views on this approach in the document ranking framework, the first one which we call the U-statistics View assumes that documents are i.i.d. random variables, while the second one which we call the Average View assumes that document pairs are i.i.d. random variables. Both views are valid in certain conditions. For example, when the relevance degree of each document is used as the ground truth, the U-statistics view is more reasonable. However, if the pairwise preferences between documents are given as the ground-truth label, it might be more reasonable to take the average view.

(1) The U-statistics View   In the U-statistics view, suppose that (x_u, y_u) and (x_v, y_v) are i.i.d. random variables according to distribution P, where $X \in \mathcal{X}$ stands for the document and $Y \in \mathcal{Y}$ stands for the ground-truth label of the document. (x_u, y_u) and (x_v, y_v) construct a pair. Given the scoring function f, a loss occurs if the documents are not ranked according to their ground truth labels. Suppose the loss function is $L(f; x_u, x_v, y_{u,v})$, where $y_{u,v} = 2 \cdot I_{\{y_u \succ y_v\}} - 1$. Again this loss function can represent pairwise 0–1 loss, or pairwise surrogate loss functions used by different algorithms. Then the *expected risk* is defined as

$$R(f) = \int_{(\mathcal{X} \times \mathcal{Y})^2} L(f; x_u, x_v, y_{u,v}) \, P(dx_u, dy_u) \, P(dx_v, dy_v). \qquad (16.3)$$

The expected risk means the loss that a ranking model f would make for two random documents. Again, it is almost impossible to compute the expected risk, and the empirical risk on the training set is used as an estimate of the expected risk. Given the training data, the *empirical risk* can be defined with the following U-statistics:

$$\hat{R}(f) = \frac{2}{m(m-1)} \sum_{u=1}^{m} \sum_{v=u+1}^{m} L(f; x_u, x_v, y_{u,v}). \qquad (16.4)$$

Specifically, when the ground truth is given as a binary relevance degree, the positive example is denoted as x^+ according to P^+ and the negative example is denoted as x^- according to P^-. Given the training data $\mathbf{x}^+ = \{x_j^+\}_{j=1}^{m^+}$ and $\mathbf{x}^- = \{x_j^-\}_{j=1}^{m^-}$ (where m^+ and m^- are the numbers of positive and negative examples in the training data respectively), the *expected risk* and the *empirical risk* can be refined as follows:

$$R(f) = \int_{\mathcal{X}^2} L(f; x^+, x^-) P^+(dx^+) P^-(dx^-), \qquad (16.5)$$

$$\hat{R}(f) = \frac{1}{m^+ m^-} \sum_{u=1}^{m^+} \sum_{v=1}^{m^-} L(f; x^+, x^-). \qquad (16.6)$$

In this case, we usually call the problem a bipartite ranking problem.

(2) The Average View   The average view assumes the i.i.d. distribution of document pairs. More specifically, with the average view, each document pair (x_u, x_v) is given a ground-truth label $y_{u,v} \in \mathcal{Y} = \{-1, 1\}$, where $y_{u,v} = 1$ indicates that document x_u is more relevant than x_v and $y_{u,v} = -1$ otherwise. Then $(x_u, x_v, y_{u,v})$ is assumed to be a random variable with probabilistic distribution P, and the expected risk can be defined as

$$R(f) = \int_{\mathcal{X}^2 \times \mathcal{Y}} L(f; x_u, x_v, y_{u,v}) P(dx_u, dx_v, dy_{u,v}). \qquad (16.7)$$

The expected risk means the loss that a ranking model f would make for a random document pair. As the distribution P is unknown, the average of the loss over $\tilde{m}$ training document pairs is used to estimate the expected risk,

$$\hat{R}(f) = \frac{1}{\tilde{m}} \sum_{j=1}^{\tilde{m}} L(f; x_{j_1}, x_{j_2}, y_{j_1, j_2}). \tag{16.8}$$

Note that the "average view" is also technically sound in certain situations. The intuition is not always right that two document pairs cannot be independent of each other when they share a common document. The reason is that the dependence (or independence) is actually defined with regards to random variables but not their values. Therefore, as long as two document pairs are sampled and labeled in an independent manner, they are i.i.d. random variables no matter whether their values (the specific documents in the pair) have overlap or not.

16.1.3 The Listwise Approach

The document ranking framework cannot describe the listwise approach. Most existing listwise ranking algorithms assume that the training set contains a deterministic set of documents associated with each query, and there is no sampling of documents. In contrast, there is no concept of a query in the document ranking framework while the sampling of documents is assumed.

16.2 Subset Ranking Framework

In the framework of subset ranking [6, 8], it is assumed that there is a hierarchical structure in the data, i.e., queries and documents associated with each query. However, only the queries are regarded as i.i.d. random variables, while the documents associated with each query is regarded as deterministically generated. For example, in [6], it is assumed that an existing search engine is used to generate the training and test data. Queries are randomly sampled from the query space. After a query is selected, the query is submitted to a search engine and the top-k documents returned will be regarded as the associated documents. In other words, there is no i.i.d. sampling with regards to documents, and each query is represented by a fixed set of documents (denoted by $\mathbf{x}$) and their ground-truth labels.

Note that generally speaking the number of documents m can be a random variable, however, for ease of discussion, here we assume it to be a fixed number for all queries.

16.2.1 The Pointwise Approach

We denote all the m documents associated with query q as $\mathbf{x} = \{x_j\}_{j=1}^m$, and their relevance degrees as $\mathbf{y} = \{y_j\}_{j=1}^m$. Note that in the subset ranking framework, there is no assumption of sampling of each individual documents. Instead, it is $(\mathbf{x}, \mathbf{y})$ (which is a representation for the corresponding query) that is regarded as a random variable sampled from the space $\mathcal{X}^m \times \mathcal{Y}^m$ according to an unknown probability distribution P.

Suppose the pointwise loss function is $L(f; x_j, y_j)$. Then the expected risk can be represented as follows,

$$R(f) = \int_{\mathcal{X}^m \times \mathcal{Y}^m} \frac{1}{m} \sum_{j=1}^m L(f; x_j, y_j) P(d\mathbf{x}, d\mathbf{y}). \qquad (16.9)$$

Intuitively, the expected risk means the *average* loss that a ranking model f would make for all the documents associated with a random query q. Since it is almost impossible to compute the expected risk, in practice, the empirical risk on the training set is used as an estimate of the expected risk.

$$\hat{R}(f) = \frac{1}{n} \sum_{i=1}^n \frac{1}{m} \sum_{j=1}^m L\big(f; x_j^{(i)}, y_j^{(i)}\big). \qquad (16.10)$$

16.2.2 The Pairwise Approach

For the pairwise approach, once again, we denote all the m documents associated with query q as $\mathbf{x} = \{x_j\}_{j=1}^m$, and denote the relevance degrees as $\mathbf{y} = \{y_j\}_{j=1}^m$. We regard $(\mathbf{x}, \mathbf{y})$ as a random variable sampled from the space $\mathcal{X}^m \times \mathcal{Y}^m$ according to an unknown probability distribution P.

Suppose the pairwise loss function is $L(f; x_v, x_v, y_{u,v})$. For any two different documents x_u and x_v, we denote $y_{u,v} = 2 \cdot I_{\{y_u \succ y_v\}} - 1$. Accordingly, the expected risk can be represented as follows,

$$R(f) = \int_{\mathcal{X}^m \times \mathcal{Y}^m} \frac{2}{m(m-1)} \sum_{u=1}^m \sum_{v=u+1}^m L(f; x_u, x_v, y_{u,v}) P(d\mathbf{x}, d\mathbf{y}). \quad (16.11)$$

Intuitively, the expected risk means the *average* loss that a ranking model f would make for all the document pairs associated with a random query q. Since it is almost impossible to compute the expected risk, in practice, the empirical risk on the training set is used as an estimate of the expected risk. In particular, given the training data $\{(\mathbf{x}^{(i)}, \mathbf{y}^{(i)})\}_{i=1}^n$, the *empirical risk* can be defined as follows,

$$\hat{R}(f) = \frac{1}{n} \sum_{i=1}^n \frac{2}{m(m-1)} \sum_{u=1}^m \sum_{v=u+1}^m L\big(f; x_u^{(i)}, x_v^{(i)}, y_{u,v}^{(i)}\big). \qquad (16.12)$$

16.2.3 The Listwise Approach

For the listwise approach, let $\mathcal{Y}$ be the output space, whose elements are permutations of m documents, denoted as π_y. Then $(\mathbf{x}, \pi_y)$ can be regarded as a random variable sampled from the space $\mathcal{X}^m \times \mathcal{Y}$ according to an unknown probability distribution P.

Suppose the listwise loss function is $L(f; \mathbf{x}, \pi_y)$. Then the expected risk can be represented as

$$R(f) = \int_{\mathcal{X}^m \times \mathcal{Y}} L(f; \mathbf{x}, \pi_y) P(d\mathbf{x}, d\pi_y). \tag{16.13}$$

Intuitively, the expected risk means the loss that a ranking model f would make for all the m documents associated with a random query q. Since it is almost impossible to compute the expected risk, in practice, the empirical risk on the training set is used as an estimate of the expected risk. In particular, given the training data $\{(\mathbf{x}^{(i)}, \pi_y^{(i)})\}_{i=1}^n$, the *empirical risk* can be defined as follows:

$$\hat{R}(f) = \frac{1}{n} \sum_{i=1}^n L\big(f; \mathbf{x}^{(i)}, \pi_y^{(i)}\big). \tag{16.14}$$

16.3 Two-Layer Ranking Framework

As pointed out in [9] and [4], the aforementioned two frameworks have their limitations in analyzing the ranking algorithms for information retrieval. The document ranking framework ignores the existence of queries, while the subset ranking framework ignores the sampling of documents. In contract, it is possible to sample both more queries and more documents to label in real information retrieval scenarios. Therefore, one should consider both queries and documents as random variables and investigate the theoretical properties when the numbers of both queries and documents approach infinity. This is exactly the motivation of the two-layer ranking framework.

16.3.1 The Pointwise Approach

Let $\mathcal{Q}$ be the query space. Each query q is assumed to be a random variable sampled from the query space with an unknown probability distribution $P_{\mathcal{Q}}$. Given this query, (x_j, y_j) is assumed to be a random variable sampled according to probability distribution $\mathcal{D}_q$ (which is dependent on query q).

Suppose the pointwise loss function is $L(f; x_j, y_j)$. Then the expected risk is defined as follows:

$$R(f) = \int_Q \int_{\mathcal{X} \times \mathcal{Y}} L(f; x_j, y_j) \mathcal{D}_q(dx_j, dy_j) P_Q(dq). \qquad (16.15)$$

Intuitively, the expected risk means the loss that a ranking model f would make for a random document associated with a random query. As both the distributions P_Q and $\mathcal{D}_q$ are unknown, the average of the loss over a set of training queries $\{q_i\}_{i=1}^n$ (i.i.d. observations according to P_Q) and their associated training documents $\{(x_j, y_j)\}_{j=1}^{m^{(i)}}$ (i.i.d. observations according to $\mathcal{D}_q$) is used to estimate the above expected risk,

$$\hat{R}(f) = \frac{1}{n} \sum_{i=1}^n \frac{1}{m^{(i)}} \sum_{j=1}^{m^{(i)}} L(f; x_j^{(i)}, y_j^{(i)}). \qquad (16.16)$$

16.3.2 The Pairwise Approach

Similar to the document ranking framework, there are also two views on the pairwise approach in the two-layer ranking framework.

(1) The U-statistics View With the U-statistics view, one assumes i.i.d. distribution of the documents and their ground-truth labels with respect to a query. Given two documents associated with query q and their ground truth labels, (x_u, y_u) and (x_v, y_v), we denote $y_{u,v} = 2 \cdot I_{\{y_u \succ y_v\}} - 1$. Then the expected risk can be defined as

$$R(f) = \int_Q \int_{(\mathcal{X} \times \mathcal{Y})^2} L(f; x_u, x_v, y_{u,v}) \mathcal{D}_q(dx_u, dy_u) \mathcal{D}_q(dx_v, dy_v) P_Q(dq).$$
$$(16.17)$$

Intuitively, the expected risk means the loss that a ranking model f would make for two random documents associated with a random query q. As both the distributions P_Q and $\mathcal{D}_q$ are unknown, the following empirical risk is used to estimate $R(f)$:

$$\hat{R}(f) = \frac{1}{n} \sum_{i=1}^n \frac{2}{m^{(i)}(m^{(i)} - 1)} \sum_{u=1}^{m^{(i)}} \sum_{v=u+1}^{m^{(i)}} L(f; x_u^{(i)}, x_v^{(i)}, y_{u,v}^{(i)}). \qquad (16.18)$$

(2) The Average View The average view assumes the i.i.d. distribution of document pairs. More specifically, with the average view, each document pair (x_u, x_v) is given a ground-truth label $y_{u,v} \in \mathcal{Y} = \{-1, 1\}$, where $y_{u,v} = 1$ indicates that document x_u is more relevant than x_v and $y_{u,v} = -1$ otherwise. Then $(x_u, x_v, y_{u,v})$ is assumed to be a random variable with probabilistic distribution $\mathcal{D}'_q$, and $R(f)$ can be defined as follows:

$$R(f) = \int_{\mathcal{Q}} \int_{\mathcal{X}^2 \times \mathcal{Y}} L(f; x_u, x_v, y_{u,v}) \, \mathcal{D}'_q(dx_u, dx_v, dy_{u,v}) P_{\mathcal{Q}}(dq). \quad (16.19)$$

As both the distributions $P_{\mathcal{Q}}$ and $\mathcal{D}'_q$ are unknown, the following empirical risk is used to estimate $R(f)$:

$$\hat{R}(f) = \frac{1}{n} \sum_{i=1}^{n} \frac{1}{\tilde{m}^{(i)}} \sum_{j=1}^{\tilde{m}^{(i)}} L(f; x_{j_1}^{(i)}, x_{j_2}^{(i)}, y_{j_1,j_2}^{(i)}). \quad (16.20)$$

16.3.3 The Listwise Approach

Note that most existing listwise ranking algorithms assume that the listwise loss function takes all the documents associated with a query as input, and there is no sampling of these documents. Therefore, the two-layer ranking framework does not explain the existing listwise ranking methods in a straightforward manner. Some modifications need to be conducted to the algorithms in order to fit them into the framework. For simplicity, we will not discuss the marriage between the two-layer ranking framework and the listwise approach in this book.

16.4 Summary

In this chapter, we have introduced three major statistical ranking frameworks used in the literature. The document ranking framework assumes the i.i.d. distribution of documents, regardless of the queries they belong to. The subset ranking framework ignores the sampling of documents per query and directly assumes the i.i.d. distribution of queries. The two-layer ranking framework considers the i.i.d. sampling of both queries and documents per query. It is clear that the two-layer ranking framework describes the real ranking problems in a more natural way. However, the other two frameworks can also be used to obtain certain theoretical results that can explain the behaviors of existing learning-to-rank methods. With the three frameworks, we give the definitions of the empirical and expected risks for different approaches to learning to rank. These definitions will be used intensively in the following two chapters, which are concerned with the generalization ability and statistical consistency of ranking methods.

16.5 Exercises

16.1 Compare the different probabilistic assumptions of the three ranking frameworks.

16.2 As mentioned in this chapter, the existing listwise ranking algorithms cannot fit into the two-layer ranking framework. Show how to modify these algorithms in order to leverage the two-layer ranking framework to analyze their theoretical properties.

References

1. Agarwal, S.: Generalization bounds for some ordinal regression algorithms. In: Proceedings of the 19th International Conference on Algorithmic Learning Theory (ALT 2008), pp. 7–21 (2008)
2. Agarwal, S., Graepel, T., Herbrich, R., Har-Peled, S., Roth, D.: Generalization bounds for the area under the roc curve. Journal of Machine Learning 6, 393–425 (2005)
3. Agarwal, S., Niyogi, P.: Stability and generalization of bipartite ranking algorithms. In: Proceedings of the 18th Annual Conference on Learning Theory (COLT 2005), pp. 32–47 (2005)
4. Chen, W., Liu, T.Y., Ma, Z.M.: Two-layer generalization analysis for ranking using rademacher average. In: Lafferty, J., Williams, C.K.I., Shawe-Taylor, J., Zemel, R., Culotta, A. (eds.) Advances in Neural Information Processing Systems 23 (NIPS 2010), pp. 370–378 (2011)
5. Clemencon, S., Lugosi, G., Vayatis, N.: Ranking and empirical minimization of U-statistics. The Annals of Statistics 36(2), 844–874 (2008)
6. Cossock, D., Zhang, T.: Subset ranking using regression. In: Proceedings of the 19th Annual Conference on Learning Theory (COLT 2006), pp. 605–619 (2006)
7. Freund, Y., Iyer, R., Schapire, R., Singer, Y.: An efficient boosting algorithm for combining preferences. Journal of Machine Learning Research 4, 933–969 (2003)
8. Lan, Y., Liu, T.Y.: Generalization analysis of listwise learning-to-rank algorithms. In: Proceedings of the 26th International Conference on Machine Learning (ICML 2009), pp. 577–584 (2009)
9. Lan, Y., Liu, T.Y., Qin, T., Ma, Z., Li, H.: Query-level stability and generalization in learning to rank. In: Proceedings of the 25th International Conference on Machine Learning (ICML 2008), pp. 512–519 (2008)
10. Rajaram, S., Agarwal, S.: Generalization bounds for k-partite ranking. In: NIPS 2005 Workshop on Learning to Rank (2005)

Chapter 17
Generalization Analysis for Ranking

Abstract In this chapter, we introduce the generalization analysis on learning-to-rank methods. In particular, we first introduce the uniform generalization bounds and then the algorithm-dependent generalization bounds. The uniform bounds hold for any ranking function in a given function class. The algorithm-dependent bounds instead consider the specific ranking function learned by the given algorithm, thus can usually be tighter. The bounds introduced in this chapter are derived under different ranking frameworks, and can explain behaviors of different learning-to-rank algorithms. We also show the limitations of existing analyses and discuss how to improve them in future work.

17.1 Overview

Generalization ability is an important theoretical property of a machine learning algorithm. It basically describes how a model learned from the training set will perform on the unseen test data. This performance is usually determined by the number of instances in the training data, and the complexity of the model. Generalization analysis has been well studied for classification, and a lot of work has been done on the topic. Comparatively, generalization analysis for ranking is not that mature, but still several attempts have been made.

There are in general two kinds of generalization analysis. The first one is called uniform generalization analysis, which tries to reveal a bound of the generalization ability for any function in the function class under investigation. In other words, such an analysis is not only applicable to the optimal model learned by a given algorithm. The second one is called algorithm-dependent generalization analysis, which is only applicable to the model learned from the training data using a specific algorithm. Since more information has been used in the second type of generalization analysis, the generalization bound is usually tighter than the uniform bound. However, as a trade off, its application scope will be smaller than that of the uniform bound.

Both types of generalization analyses heavily depend on the statistical ranking frameworks introduced in the previous chapter. For example, with the document ranking framework, one is concerned with the generalization over documents (or document pairs); with the subset ranking framework, one is concerned with the

T.-Y. Liu, *Learning to Rank for Information Retrieval*,
DOI 10.1007/978-3-642-14267-3_17, © Springer-Verlag Berlin Heidelberg 2011

generalization over queries; and with the two-layer ranking framework, one is concerned with the generalization over both queries and documents.

In this chapter, we will introduce existing works on generalization analysis for ranking, with respect to different types of bounds and different ranking frameworks.

17.2 Uniform Generalization Bounds for Ranking

Several works have been done on the uniform generalization bounds of either the true loss or the surrogate loss functions in ranking. Some of them fall into the document ranking framework, while others fall into the subset ranking framework and the two-layer ranking framework.

17.2.1 For Document Ranking

In [10], the uniform generalization ability for RankBoost is discussed in the context of bipartite document ranking. That is, there are two levels of relevance degrees in the data and there is no notion of query considered in the analysis. The generalization bound is defined with respect to the pairwise 0–1 loss. That is, if a pair of documents (whose labels are positive and negative respectively) is correctly ranked, the loss is zero, otherwise it is one. The corresponding empirical and expected risks are defined as in (16.5) and (16.6) (see Chap. 16).

To perform the analysis, the VC dimension [16, 17] is used. According to the theoretical results obtained in [10] (see the following theorem), the generalization bound converges to zero at a rate of $O(\max\{\sqrt{\frac{\log(m^+)}{m^+}}, \sqrt{\frac{\log(m^-)}{m^-}}\})$, where m^+ and m^- are the numbers of relevant and irrelevant documents respectively.

Theorem 17.1 *Assume that all the weak learners belong to function class $\mathcal{F}'$, which has a finite VC dimension V, the scoring functions f (which are the weighted combinations of the weak rankers) belong to function class $\mathcal{F}$. Let S^+ and S^- be positive and negative samples of size m^+ and m^-, respectively. That is, $S^+ = \{x_j^+\}_{j=1}^{m^+}$ and $S^+ = \{x_j^-\}_{j=1}^{m^-}$. Then with probability at least $1 - \delta$ ($0 < \delta < 1$), the following inequality holds for $\forall f \in \mathcal{F}$:*

$$|R_0(f) - \hat{R}_0(f)| \leq 2\sqrt{\frac{V'(\log \frac{2m^+}{V'} + 1) + \log \frac{18}{\delta}}{m^+}}$$

$$+ 2\sqrt{\frac{V'(\log \frac{2m^-}{V'} + 1) + \log \frac{18}{\delta}}{m^-}}, \qquad (17.1)$$

where $V' = 2(V + 1)(T + 1)\log_2(e(T + 1))$, T is the number of weak rankers in RankBoost.

While the above work uses the concept of the VC dimension for general-ization analysis, the notion of the bipartite rank-shatter coefficient, denoted as $r(\mathcal{F}, m^+, m^-)$, is used in [2]. This new notion has a similar meaning to the shatter-ing coefficient in classification [16, 17].

Using the bipartite rank-shatter coefficient as a tool, the following theorem has been proven in [2]. For the class of linear scoring functions in the one-dimensional feature space, $r(\mathcal{F}, m^+, m^-)$ is a constant, regardless of the values of m^+ and m^-. In this case, the bound converges to zero at a rate of $O(\max\{\frac{1}{\sqrt{m^+}}, \frac{1}{\sqrt{m^-}}\})$, and is therefore tighter than the bound based on the VC dimension (as given in Theorem 17.1). For the class of linear scoring functions in the d-dimensional fea-ture space $(d > 1)$, $r(\mathcal{F}, m^+, m^-)$ is of the order $O((m^+ m^-)^d)$, and in this case the bound has a similar convergence rate to that based on the VC dimension, i.e., $O(\max\{\sqrt{\frac{\log(m^+)}{m^+}}, \sqrt{\frac{\log(m^-)}{m^-}}\})$.

Theorem 17.2 *Let $\mathcal{F}$ be the class of real-valued functions on $\mathcal{X}$, then with prob-ability at least $1 - \delta$ $(0 < \delta < 1)$,*

$$\forall f \in \mathcal{F}, \quad |R_0(f) - \hat{R}_0(f)| \le \sqrt{\frac{8(m^+ + m^-)(\log\frac{4}{\delta} + \log r(\mathcal{F}, 2m^+, 2m^-))}{m^+ m^-}}.$$

$$(17.2)$$

Note that the above two results are only applicable to the case of bipartite ranking. Some other researchers have extended these results to more general cases, e.g., the case of k-partite ranking [9, 15].

Here we take [9] as an example. In this work, the labels of documents are no longer assumed to be positive or negative. Instead, a document is preferred to an-other one if and only if the label of the former document is larger than that of the latter document. Specifically, given two documents x_u and x_v whose labels are y_u and y_v, the pairwise 0–1 loss is defined as $I_{\{(y_u - y_v)(f(x_u) - f(x_v)) < 0\}}$. In other words, if the ranking result given by function f is in the same order of that given by the ground-truth label, the loss is zero; otherwise the loss is one. Based on the pairwise 0–1 loss, the expected risks and empirical risks are defined as below.

$$R_0(f) = E[I_{\{(y_u - y_v)(f(x_u) - f(x_v)) < 0\}}], \qquad (17.3)$$

$$\hat{R}_0(f) = \frac{2}{m(m-1)} \sum_{u=1}^{m} \sum_{v=u+1}^{m} I_{\{(y_u - y_v)(f(x_u) - f(x_v)) < 0\}}. \qquad (17.4)$$

Using the properties of U-statistics [9], the following generalization bound is obtained with respect to the aforementioned empirical and expected risks.

Theorem 17.3 *With probability at least* $1 - \delta$ $(0 < \delta < 1)$, *the following inequality holds:*

$$\left| R_0(f) - \hat{R}_0(f) \right| \leq 4c \sqrt{\frac{V(\hat{\mathcal{F}})}{m}} + 4 \sqrt{\frac{\ln \frac{1}{\delta}}{m-1}}, \tag{17.5}$$

where $\hat{\mathcal{F}} = \{ \text{sgn}(f(x_u) - f(x_v)); f \in \mathcal{F} \}$, *and* $V(\hat{\mathcal{F}})$ *is its VC dimension; c is a universal constant.*

As compared to the results in [10], the bound as shown in Theorem 17.3 seems tighter, although it is also based on the VC dimension. This is mainly due to the consideration of the properties of U-statistics in the analysis.

In addition to the discussions on the pairwise 0–1 loss, in [9], the generalization bounds with respect to surrogate loss functions are also discussed. Suppose the surrogate loss function is ϕ, then the expected and empirical surrogate risks can be defined as below.

$$R_\phi(f) = E\left[\phi\left(-\text{sgn}(y_u - y_v) \cdot \left(f(x_u) - f(x_v) \right) \right) \right], \tag{17.6}$$

$$\hat{R}_\phi(f) = \frac{2}{m(m-1)} \sum_{u=1}^{m} \sum_{v=u+1}^{m} \phi\left(-\text{sgn}(y_u - y_v) \cdot \left(f(x_u) - f(x_v) \right) \right). \tag{17.7}$$

The corresponding generalization bound is shown in the following theorem.

Theorem 17.4 *With probability at least* $1 - \delta$ $(0 < \delta < 1)$, *the following inequality holds:*

$$\left| R_\phi(f) - \hat{R}_\phi(f) \right| \leq 4cB\phi'(B) \frac{V}{\sqrt{m}} + \sqrt{\frac{B^2 \ln \frac{1}{\delta}}{2m}}. \tag{17.8}$$

Here $\hat{\mathcal{F}} = \{ \text{sgn}(f(x_u) - f(x_v)); f \in \mathcal{F} \}$, *and* $V(\hat{\mathcal{F}})$ *is its VC dimension; c is a universal constant; the output of function f is uniformly bounded by* $-B$ *and* B.

17.2.2 For Subset Ranking

In [13], the uniform bounds for the listwise surrogate loss functions are obtained in the setting of subset ranking. Since queries are considered i.i.d. random variables in subset ranking, the corresponding generalization bound is also referred to as the query-level generalization bound. Under the subset ranking framework, the expected and empirical risks are defined as in (16.13) and (16.14) (see Chap. 16).

Technically, the theory of the Rademacher average (RA) [4, 5] is used in the analysis. RA measures how much the function class can fit random noise, which is defined below.

Definition 17.1 For a function class $\mathscr{G}$, the empirical RA is defined as

$$\widehat{\mathscr{R}}(\mathscr{G}) = E_\sigma \sup_{g \in \mathscr{G}} \frac{1}{n} \sum_{i=1}^{n} \sigma_i g(z_i), \tag{17.9}$$

where $z_i, i = 1, \ldots, n$ are i.i.d. random variables, and $\sigma_i, i = 1, \ldots, n$ are i.i.d. random variables, with probability $\frac{1}{2}$ of taking a value of $+1$ or -1.

Based on the RA theory [4, 5], the following generalization bounds have been derived. Here it is assumed that $\forall x \in \mathscr{X}, \|x\| \leq M$, and the scoring function f is learned from the linear function class $\mathscr{F} = \{x \rightarrow w^T x : \|w\| \leq B.\}$, for simplicity. In this case, one has $\forall x \in \mathscr{X}, |f(x)| \leq BM$.

Theorem 17.5 *Let $\mathscr{A}$ denote ListNet or ListMLE, and let $L_\mathscr{A}(f; \mathbf{x}, \pi_y)$ be the corresponding listwise loss. Given the training data $S = \{(\mathbf{x}^{(i)}, \pi_y^{(i)}), i = 1, \ldots, n\}$, $\forall f \in \mathscr{F}, (\mathbf{x}, \pi_y) \in \mathscr{X}^m \times \mathscr{Y}, L_\mathscr{A}(f; \mathbf{x}, \pi_y) \in [0, 1]$, with probability at least $1 - \delta$ $(0 < \delta < 1)$, the following inequality holds:*

$$\sup_{f \in \mathscr{F}} \left(R_\phi(f) - \widehat{R}_\phi(f) \right) \leq 2C_\mathscr{A}(\varphi) N(\varphi) \widehat{\mathscr{R}}(\mathscr{F}) + \sqrt{\frac{2\log\frac{2}{\delta}}{n}}, \tag{17.10}$$

where $\widehat{\mathscr{R}}(\mathscr{F})$ is the RA of the scoring function class (for the linear scoring function, we have $\widehat{\mathscr{R}}(\mathscr{F}) \leq \frac{2BM}{\sqrt{n}}$); $N(\varphi) = \sup_{x \in [-BM, BM]} \varphi'(x)$ measures the smoothness of the transformation function φ; $C_\mathscr{A}(\varphi)$ is an algorithm-dependent factor.

The expressions of $N(\varphi)$ and $C_\mathscr{A}(\varphi)$ for ListNet and ListMLE, with respect to three representative transformation functions are listed in Table 17.1.[1]

From Theorem 17.5, one can see that when the number of training queries n approaches infinity, the query-level generalization bound will converge to zero at a rate of $O(\frac{1}{\sqrt{n}})$. Furthermore, by comparing the query-level generalization bound for different listwise ranking algorithms, and with regards to different transformation functions, one can have the following observations.

- The query-level generalization bound for ListMLE is much tighter than that for ListNet, especially when m, the length of the list, is large.
- The query-level generalization bound for ListMLE decreases monotonously, while that of ListNet increases monotonously, with respect to m.
- The linear transformation function is the best choice in terms of the query-level generalization bound in most cases.

---

[1] The three transformation functions are

◇ Linear Functions: $\varphi_L(x) = ax + b, x \in [-BM, BM]$.
◇ Exponential Functions: $\varphi_E(x) = e^{ax}, x \in [-BM, BM]$.
◇ Sigmoid Functions: $\varphi_S(x) = \frac{1}{1+e^{-ax}}, x \in [-BM, BM]$.

Table 17.1 $N(\varphi)$ and $C_{\mathscr{A}}(\varphi)$ for ListNet and ListMLE

φ	$N(\varphi)$	$C_{\text{ListMLE}}(\varphi)$	$C_{\text{ListNet}}(\varphi)$
φ_L	a	$\dfrac{2}{(b-aBM)(\log m + \log\frac{b+aBM}{b-aBM})}$	$\dfrac{2m!}{(b-aBM)(\log m + \log\frac{b+aBM}{b-aBM})}$
φ_E	ae^{aBM}	$\dfrac{2e^{aBM}}{\log m + 2aBM}$	$\dfrac{2m!e^{aBM}}{\log m + 2aBM}$
φ_S	$\dfrac{ae^{aBM}}{(1+e^{-aBM})^2}$	$\dfrac{2(1+e^{aBM})}{\log m + aBM}$	$\dfrac{2m!(1+e^{aBM})}{\log m + aBM}$

17.2.3 For Two-Layer Ranking

As can be seen, the generalization bounds introduced in the previous subsections are over either documents or queries. This is mainly because of the statistical frameworks that they have used. As a result, they cannot explain some empirical observations. For example, the experiments in [18] show that given a fixed total number of judgments, in order to achieve the best test performance, neither extremely deep labeling (i.e., labeling a large number of queries and only a few documents per query) nor extremely shallow labeling (i.e., labeling a few queries but a large number of documents per query) is a good choice. This is clearly not in accordance with any of the above generalization bounds.

To address the aforementioned issues, in [8], generalization analysis for the pairwise ranking algorithms is conducted, under the two-layer ranking framework. The expected and empirical risks are defined as in (16.17) and (16.18) (see Chap. 16).

In particular, the following result has been obtained in [8], also based on the concept of Rademacher average (RA) [4, 5].

Theorem 17.6 *Suppose L is the loss function for pairwise ranking. Assume (1) $L \circ \mathscr{F}$ is bounded by M, (2) $\mathrm{E}[\mathscr{R}_m(L \circ \mathscr{F})] \leq D(L \circ \mathscr{F}, m)$, then with probability at least $1 - \delta$ $(0 < \delta < 1)$, $\forall f \in \mathscr{F}$*

$$R(f) \leq \hat{R}(f) + D(L \circ \mathscr{F}, n) + \sqrt{\frac{2M^2 \log(\frac{4}{\delta})}{n}} + \frac{1}{n}\sum_{i=1}^{n} D\left(L \circ \mathscr{F}, \left\lfloor \frac{m^{(i)}}{2} \right\rfloor \right)$$

$$+ \sqrt{\sum_{i=1}^{n} \frac{2M^2 \log\frac{4}{\delta}}{m^{(i)}n^2}},$$

where for certain ranking function class, $D(L \circ \mathscr{F}, m)$ has its explicit form. For example, for the function class $\tilde{\mathscr{F}} = \{f(x, x') = f(x) - f(x'); f \in \mathscr{F}\}$ whose VC dimension is V, and $|f(x)| \leq B$, it has been proven that $D(L_0 \circ \mathscr{F}, m) = c_1\sqrt{V/m}$ and $D(L_\phi \circ \mathscr{F}, m) = c_2 B\phi'(B)\sqrt{V/m}$ in [4, 5], where c_1 and c_2 are both constants.

According to the above theorem, one may have the following discussions.

- The increasing number of either queries or documents per query in the training data will enhance the two-layer generalization ability.
- Only if $n \to \infty$ and $m^{(i)} \to \infty$ simultaneously does the two-layer generalization bound uniformly converge.

- If we only have a limited budget to label C documents in total, according to Theorem 17.6, there is an optimal trade off between the number of training queries and that of training documents per query. This is consistent with previous empirical findings in [18].

Actually one can attain the optimal trade off by solving the following optimization problem:

$$\min_{n, m^{(1)}, \ldots, m^{(n)}} \quad D(L \circ \mathscr{F}, n) + \sqrt{\frac{2M^2 \log(\frac{4}{\delta})}{n}}$$

$$+ \frac{1}{n} \sum_{i=1}^{n} D\left(L \circ \mathscr{F}, \left\lfloor \frac{m^{(i)}}{2} \right\rfloor\right) + \sqrt{\sum_{i=1}^{n} \frac{2M^2 \log \frac{4}{\delta}}{m^{(i)} n^2}}$$

$$\text{s.t.} \quad \sum_{i=1}^{n} m_i = C.$$

This optimization problem is easy to solve. For example, if $VC(\tilde{\mathscr{F}}) = V$, for the pairwise 0–1 loss, the solution to the optimization problem is $n^* = \frac{c_1 \sqrt{V} + \sqrt{2 \log(4/\delta)}}{c_1 \sqrt{2V}} \sqrt{C}$, $m^{(i)*} \equiv \frac{C}{n^*}$, where c_1 is a constant. From this result we note the following.

- n^* decreases with the increasing capacity of the function class. That is, we should label fewer queries and more documents per query when the hypothesis space is larger.
- For fixed hypothesis space, n^* increases with the confidence level δ. That is, we should label more query if we want the bound to hold with a larger probability.

The above findings can be used to explain the behavior of existing pairwise ranking algorithms, and can be used to guide the construction of training set for learning to rank.

17.3 Algorithm-Dependent Generalization Bound

Note that the uniform generalization bound is valid for any function in the hypothesis space under investigation, no matter whether the function is the optimal one learned by a given algorithm or not. Therefore, it is understandable that the bound might not be very tight. In contrast, it is possible that we can get a tighter bound for the optimal function learned by the algorithm. This is just the motivation of developing the algorithm-dependent generalization bounds.[2] Again, in this section, we

---

[2]Note that the disadvantage of algorithm-dependent bounds lies in that they can only be used for specific algorithms, and may not be derived for every algorithm.

will introduce previous work according to the statistical ranking frameworks that they use.

17.3.1 For Document Ranking

In [1, 3], the *stability* is used as a tool to derive the algorithm-dependent generalization bound for pairwise ranking algorithms, under the framework of document ranking.

Let $\mathscr{A}$ be a ranking algorithm, and let L be the loss function that is minimized in $\mathscr{A}$. Suppose we have learned a ranking model f_1 from m training documents using algorithm $\mathscr{A}$. Then we replace a training document with another document and learn a new model f_2 from the new training data. We say that $\mathscr{A}$ has uniform loss stability β with respect to L, if the difference between the losses with respect to f_1 and f_2 on any unseen document pair (x_u, x_v) is smaller than $\beta(m)$. In other words, if the model learned from the training data is robust to the small change in the training data, the algorithm is regarded as having certain stability.

The generalization bound obtained in [1, 3] is as shown in Theorem 17.7.

Theorem 17.7 *Let L be the loss function with $L(f; x_u, x_v, y_{u,v}) \in [0, B]$, and $\mathscr{A}$ has uniform loss stability β. Then with probability at least $1 - \delta$ ($0 < \delta < 1$), the following inequality holds:*

$$R(f_S) - \hat{R}(f_S) \le 2\beta(m) + \left(m\beta(m) + B\right)\sqrt{\frac{2\ln(\frac{1}{\delta})}{m}}. \qquad (17.11)$$

For different algorithms, one can get different stability coefficients (note that not all algorithms are stable, and therefore the stability coefficients of some algorithms do not approach zero when the number of training data m approaches infinity). In particular, symmetric ranking algorithms with general regularization terms are stable. For example, Ranking SVM is a stable algorithm, and its stability coefficient is

$$\beta(m) = \frac{16\kappa^2}{\lambda m}, \qquad (17.12)$$

where $\forall x \in \mathscr{X}$, $K(x, x) \le \kappa^2 < \infty$, λ is the tradeoff coefficient for the regularization term in Ranking SVM.

As can be seen from the above discussions, for algorithms like Ranking SVM [11, 12], the bound converges to zero at a rate of $O(\frac{1}{\sqrt{m}})$. This bound is tighter than the bound based on the Rank-Shatter Coefficient (see inequality (17.2)) except for the case of using linear scoring functions in the one-dimensional feature space. This clearly shows the advantage of the algorithm-dependent bound.

17.3.2 For Subset Ranking

In [14], the stability theory [6] is extended to perform query-level generalization analysis on the pairwise approach, under the framework of subset ranking. In particular, the *average view* is taken in the analysis.

To assist the analysis, the concept of uniform leave-one-query-out pairwise loss stability (also referred to as *query-level stability* for short) is proposed in [14]. Suppose we have learned a ranking model f_1 from a training set with n queries, using an algorithm $\mathscr{A}$. Suppose L is the loss function that is minimized in algorithm $\mathscr{A}$. Then we randomly remove a query and all its associated document pairs from the training data, and learn a ranking model f_2 from the new training set. If the difference between the losses with respect to f_1 and f_2 on any unseen document pair (x_u, x_v) is smaller than $\tau(n)$, we say the algorithm $\mathscr{A}$ has uniform leave-one-query-out pairwise loss stability with coefficient τ with respect to L.

Based on the concept of the query-level stability, a query-level generalization bound has been derived in [14], as shown in Theorem 17.8. The theorem states that if a pairwise ranking algorithm has query-level stability, then with a large probability, the expected query-level risk can be bounded by the empirical query-level risk and a term that depends on the query number and the stability of the algorithm.

Theorem 17.8 *Let $\mathscr{A}$ be a pairwise ranking algorithm, $(q_1, S^{(1)}), \ldots, (q_n, S^{(n)})$ be n training queries and their associated labeled documents, and let L be the pairwise loss function. If*

1. $\forall (q_1, S^{(1)}), \ldots, (q_n, S^{(n)}), \forall q \in \mathscr{Q}, (x_u, x_v, y_{u,v}) \in \mathscr{X}^2 \times \mathscr{Y}, |L(f_{(q_i, S^{(i)})_{i=1}^n};$
   $x_u, x_v, y_{u,v})| \leq B,$
2. $\mathscr{A}$ *has query-level stability with coefficient* τ,

then with probability at least $1 - \delta$ ($0 < \delta < 1$) *over the samples of* $\{(q_i, S^{(i)})\}_{i=1}^n$ *in the product space* $\prod_{i=1}^n \{\mathscr{Q} \times (\mathscr{X}^2 \times \mathscr{Y})^\infty\}$, *the following inequality holds:*

$$R(f_{\{(q_i, S^{(i)})\}_{i=1}^n}) \leq \widehat{R}(f_{\{(q_i, S^{(i)})\}_{i=1}^n}) + 2\tau(n) + (4n\tau(n) + B)\sqrt{\frac{\log \frac{1}{\delta}}{2n}}. \quad (17.13)$$

When using this theorem to perform the query-level generalization analysis on pairwise ranking algorithms, what one needs to do is to compute the query-level stability coefficient $\tau(n)$ of the algorithms.

For example, as shown in [14], Ranking SVM [11, 12] has a query-level stability with coefficient $\tau(n) = \frac{4\kappa^2}{\lambda n} \times \max \frac{\tilde{m}^{(i)}}{\frac{1}{n}\sum_{i=1}^n \tilde{m}^{(i)}}$, where $\tilde{m}^{(i)}$ is the number of document pairs associated with query q_i, and $\forall x \in \mathscr{X}, K(x, x) \leq \kappa^2 < \infty$. On this basis, one can have the following discussions regarding the query-level generalization ability of Ranking SVM.

- When the number of training queries approaches infinity, with a large probability the empirical risk of Ranking SVM will converge to its expected risk, at a rate of $O(\frac{1}{\sqrt{n}})$.

- When the number of training queries is finite, the expected risk and the empirical risk are not necessarily close to each other.

Furthermore, as shown in [14], IR-SVM [7] has a query-level stability with coefficient $\tau(n) = \frac{4\kappa^2}{\lambda n}$. On this basis, one can find that when the number of training queries approaches infinity, with a large probability the empirical risk of IR-SVM will converge to its expected risk, at a convergence rate of $O(\frac{1}{\sqrt{n}})$. When the number of queries is finite, the query-level generalization bound is a decreasing function of the number of training queries.

By comparing the query-level generalization abilities of Ranking SVM and IR-SVM, we can find that the convergence rates for these two algorithms are both $O(\frac{1}{\sqrt{n}})$. However, by comparing the case with a finite number of training queries, the bound for IR-SVM is much tighter than that for Ranking SVM.

17.3.3 For Two-Layer Ranking

As far as we know, there is still no work on the algorithm-dependent generalization bounds under the two-layer ranking framework.

17.4 Summary

In this chapter, we have introduced the generalization analysis on ranking algorithms, with different types of bounds, and under different statistical ranking frameworks.

As can be seen, early works assume the use of the document ranking framework. As a result, the corresponding theoretical results can describe the accuracy of ranking unseen document pairs. However, as mentioned in [14], such a result is not highly desired in the context of information retrieval. This is mainly because people care more about the ranking accuracy for new queries but not for new document pairs. Some attempts have been made along this direction [13, 14], however, these theoretical results can only explain the change of generalization ability with respect to the increasing number of training queries, and ignore the influence of the number of training documents per query on the generalization ability. In order to obtain such kinds of results, one needs to investigate the generalization theory under the two-layer ranking framework. Work along this line is, however, still very limited [8]. For example, currently the two-layer generalization ability of listwise ranking methods is not clear, and the study on two-layer stability (or other algorithm-dependent bounds) is still missing. More work is expected to be conducted along this line.

For clarity, we summarize what have been introduced in this chapter using Tables 17.2 and 17.3. From the tables, we can see that there are still large space to explore along this direction. Generalization analysis for ranking is still an emerging research area, and we can expect its fast development in the future.

Table 17.2 Uniform generalization bounds for ranking

Approaches	Document ranking	Subset ranking	Two-layer ranking
Pointwise	–	–	–
Pairwise	[2, 9, 10, 15]	–	[8]
Listwise	–	[13]	–

Table 17.3 Algorithm-dependent bounds for ranking

Approaches	Document ranking	Subset ranking	Two-layer ranking
Pointwise	–	–	–
Pairwise	[1, 3]	[14]	–
Listwise	–	–	–

17.5 Exercises

17.1 Two-layer learning is meaningful not only for information retrieval, but also for many other applications. Please give two example applications where there are hierarchical structures in the data, and two-layer sampling is needed to generate the training data.

17.2 Investigate whether RankBoost, RankNet, and FRank are stable with respect to their pairwise loss functions, and derive their stability coefficients if they exist.

17.3 Investigate whether ListNet and ListMLE are stable with respect to their list-wise loss functions, and derive their query-level stability coefficients if they exist.

17.4 Generalization analysis is mainly a theoretical tool. However, the result can also be empirically verified to some extent. Design an experiment to validate the results presented in Sect. 17.2.3.

References

1. Agarwal, S.: Generalization bounds for some ordinal regression algorithms. In: Proceedings of the 19th International Conference on Algorithmic Learning Theory (ALT 2008), pp. 7–21 (2008)
2. Agarwal, S., Graepel, T., Herbrich, R., Har-Peled, S., Roth, D.: Generalization bounds for the area under the roc curve. Journal of Machine Learning **6**, 393–425 (2005)
3. Agarwal, S., Niyogi, P.: Stability and generalization of bipartite ranking algorithms. In: Proceedings of the 18th Annual Conference on Learning Theory (COLT 2005), pp. 32–47 (2005)
4. Bartlett, P.L., Mendelson, S.: Rademacher and Gaussian complexities risk bounds and structural results. Journal of Machine Learning **3**, 463–482 (2003)
5. Bousquet, O., Boucheron, S., Lugosi, G.: Introduction to statistical learning theory. In: Advanced Lectures on Machine Learning, pp. 169–207. Springer, Berlin (2004)

6. Bousquet, O., Elisseeff, A.: Stability and generalization. The Journal of Machine Learning Research **2**, 449–526 (2002)
7. Cao, Y., Xu, J., Liu, T.Y., Li, H., Huang, Y., Hon, H.W.: Adapting ranking SVM to document retrieval. In: Proceedings of the 29th Annual International ACM SIGIR Conference on Research and Development in Information Retrieval (SIGIR 2006), pp. 186–193 (2006)
8. Chen, W., Liu, T.Y., Ma, Z.M.: Two-layer generalization analysis for ranking using rademacher average. In: Lafferty, J., Williams, C.K.I., Shawe-Taylor, J., Zemel, R., Culotta, A. (eds.) Advances in Neural Information Processing Systems 23 (NIPS 2010), pp. 370–378 (2011)
9. Clemencon, S., Lugosi, G., Vayatis, N.: Ranking and empirical minimization of u-statistics. The Annals of Statistics **36**(2), 844–874 (2008)
10. Freund, Y., Iyer, R., Schapire, R., Singer, Y.: An efficient boosting algorithm for combining preferences. Journal of Machine Learning Research **4**, 933–969 (2003)
11. Herbrich, R., Obermayer, K., Graepel, T.: Large margin rank boundaries for ordinal regression. In: Advances in Large Margin Classifiers, pp. 115–132 (2000)
12. Joachims, T.: Optimizing search engines using clickthrough data. In: Proceedings of the 8th ACM SIGKDD International Conference on Knowledge Discovery and Data Mining (KDD 2002), pp. 133–142 (2002)
13. Lan, Y., Liu, T.Y.: Generalization analysis of listwise learning-to-rank algorithms. In: Proceedings of the 26th International Conference on Machine Learning (ICML 2009), pp. 577–584 (2009)
14. Lan, Y., Liu, T.Y., Qin, T., Ma, Z., Li, H.: Query-level stability and generalization in learning to rank. In: Proceedings of the 25th International Conference on Machine Learning (ICML 2008), pp. 512–519 (2008)
15. Rajaram, S., Agarwal, S.: Generalization bounds for k-partite ranking. In: NIPS 2005 Workshop on Learning to Rank (2005)
16. Vapnik, V.N.: The Nature of Statistical Learning Theory. Springer, Berlin (1995)
17. Vapnik, V.N.: Statistical Learning Theory. Wiley-Interscience, New York (1998)
18. Yilmaz, E., Robertson, S.: Deep versus shallow judgments in learning to rank. In: Proceedings of the 32st Annual International Conference on Research and Development in Information Retrieval (SIGIR 2009), pp. 662–663 (2009)

Chapter 18
Statistical Consistency for Ranking

Abstract In this chapter, we introduce the statistical consistency of learning-to-rank methods. In particular, we will introduce the existing results on statistical consistency under different ranking frameworks, and with respect to different true losses, e.g., the pairwise 0–1 loss, the permutation-level 0–1 loss, the top-k loss, and the weighted Kendall's τ loss. Then we will make discussions on these results and point out the ways of further improving them.

18.1 Overview

As we can see, for the ranking algorithms introduced in this book, a common practice is to use a surrogate loss function instead of the true loss in the learning process (this is true even for some direct optimization algorithms). However, it is unclear whether the ranking function learned by minimizing the surrogate loss function can have good ranking performances in terms of the true loss function. This is just what statistical consistency is concerned about.

Specifically, when the number of training samples approaches infinity, if the expected true risk of the ranking function learned by minimizing an expected surrogate risk can converge to the minimum expected true risk, we say that the algorithm minimizing the surrogate risk is *statistically consistent*. In practice, in order to prove the statistical consistency, one usually needs to derive a regret bound. That is, if the difference between the expected true risk of a ranking function and the minimum expected true risk can be bounded by the difference between the expected surrogate risk of the ranking function and the minimum expected surrogate risk, we can come to the conclusion that the learning method that minimizes the surrogate risk is statistically consistent.

In this chapter, we will introduce some previous works that analyze the statistical consistency of ranking methods. As can be seen later, these works are different from each other, in terms of the statistical ranking frameworks and the true loss functions.

T.-Y. Liu, *Learning to Rank for Information Retrieval*,
DOI 10.1007/978-3-642-14267-3_18, © Springer-Verlag Berlin Heidelberg 2011

18.2 Consistency Analysis for Document Ranking

18.2.1 Regarding Pairwise 0–1 Loss

In [5], the consistency for pairwise ranking methods is discussed, under the framework of document ranking.

Give two documents x_u and x_v whose labels are y_u and y_v, the pairwise 0–1 loss is defined as $I_{\{(y_u-y_v)(f(x_u)-f(x_v))<0\}}$. In other words, if the ranking result given by function f is in the same order of that given by the ground-truth label, the loss is zero; otherwise the loss is one. Based on the pairwise 0–1 loss, the expected true risk is defined as (the expectation is taken over documents according to the document ranking framework)

$$R_0(f) = E[I_{\{(y_u-y_v)(f(x_u)-f(x_v))<0\}}]. \tag{18.1}$$

The expected surrogate risk is defined by

$$R_\phi(f) = E\big[\phi\big(-\mathrm{sgn}(y_u - y_v) \cdot \big(f(x_u) - f(x_v)\big)\big)\big], \tag{18.2}$$

where ϕ is a convex function, e.g., the exponential, logistic, and hinge function.

The following regret bound is derived in [5].

Theorem 18.1 *Suppose $R_0^* = \inf_f R_0(f)$ and $R_\phi^* = \inf_f R_\phi(f)$. Then for all functions f, as long as ϕ is convex, we have*

$$R_0(f) - R_0^* \le \psi^{-1}\big(R_\phi(f) - R_\phi^*\big), \tag{18.3}$$

where

$$\psi(x) = H^-\left(\frac{1+x}{2}\right) - H^-\left(\frac{1-x}{2}\right),$$

$$H^-(\rho) = \inf_{\alpha:\alpha(2\rho-1)\le 0}\big(\rho\phi(-\alpha) + (1 - \rho)\phi(\alpha)\big). \tag{18.4}$$

This theorem basically says that if an algorithm can effectively minimize the expected surrogate risk to its minimum (the right-hand side of the inequality approaches zero), then the expected true risk will also be minimized to its minimum (the left-hand side of the inequality also approaches zero). In other words, the algorithm is statistically consistent if ϕ is a convex function. According to the result, it is not difficult to verify that Ranking SVM [8, 9], RankBoost [7], and RankNet [3] are all consistent with the pairwise 0–1 loss.

18.3 Consistency Analysis for Subset Ranking

There are several works on consistency analysis under the subset ranking framework. All these works assume that the expected risks are computed over the queries.

18.3.1 Regarding DCG-Based Ranking Error

In [6], the consistency issue with respect to DCG-based ranking error is studied. In particular, the study is focused on a specific regression-based ranking method. The loss function minimized by the method has the following form:

$$L_\phi(f, \mathbf{x}, \mathbf{y}) = \sum_{j=1}^{m} w_j \big(f(x_j) - y_j\big)^2 + u \sup_j w'_j \big(\max\{0, \big(f(x_j) - \delta\big)\}\big)^2. \quad (18.5)$$

The weight function w_j is chosen so that it focuses only on the most important examples (e.g., those examples whose $y_j = 1$) while w'_j focuses on the examples not covered by w_j (e.g., those examples whose $y_j = 0$). This loss function basically requires that the output of the ranking function f on a relevant document should be as close to 1 as possible, and at the same time requires that the output on an irrelevant document should not be larger than a small positive value δ by much.

As for the minimization of the above loss function, the following result has been obtained.

Theorem 18.2 *For all scoring functions f, let $r_f = \text{sort} \circ f$ be the induced ranking model. Let f_B be the Bayes optimal scoring function, i.e., $f_B(x_j) = y_j$, and denote the corresponding ranking model as $r_B = \text{sort} \circ f_B$. Suppose f^* is the scoring function learned by minimizing $E[L_\phi(f; \mathbf{x}, \mathbf{y})]$, then in certain conditions (see [6] for more details) the following inequality holds:*

$$E[DCG(r_B, \mathbf{x}, \mathbf{y})] - E[DCG(r_f, \mathbf{x}, \mathbf{y})]$$
$$\leq C\big(E[L_\phi(f; \mathbf{x}, \mathbf{y})] - E[L_\phi(f^*; \mathbf{x}, \mathbf{y})]\big)^{1/2}, \quad (18.6)$$

where C is a constant.

The above theorem basically says that in certain conditions, the minimization of the expectation of the regression loss given in (18.5) will lead to the maximization of the expected DCG value. In other words, the regression loss in (18.5) is statistically consistent with DCG-based ranking error.

18.3.2 Regarding Permutation-Level 0–1 Loss

In [13], the consistency of listwise ranking methods is investigated, with respect to the permutation-level 0–1 loss. The permutation-level 0–1 loss takes a value of 0 if the ranked list given by the ranking function is exactly the same as the ground-truth permutation; and takes a value of 1 otherwise.

In particular, two conditions for the consistency are given in [13]. One is for the underlying probability space, called *order preserving*. The other is for the surrogate loss function, called *order sensitive*. The intuitive explanation of these two concepts are as follows.

- The property of order preserving implies the ranking of a document is inherently determined by its own. This property has been explicitly or implicitly used in many previous works. For example, in [6], it is argued that one can make the relevance score of a document independent of other documents, by choosing an appropriately defined feature function.
- The property of order sensitive shows that when starting with a ground-truth permutation, the loss will increase if we exchange the positions of two documents in it, and the speed of increase in loss is sensitive to the positions of the two documents.

Based on the above two concepts, the following theorem has been obtained.

Theorem 18.3 *Let $L_\phi(f; \mathbf{x}, \pi_y)$ be an order sensitive loss function. $\forall m$ documents, if its permutation probability space is order preserving with respect to $m - 1$ document pairs $(j_1, j_2), (j_2, j_3), \ldots, (j_m - 1, j_m)$, then the loss $L_\phi(f; \mathbf{x}, \pi_y)$ is consistent with respect to the permutation-level 0–1 loss.*

Further studies show that the surrogate loss functions in ListNet [4] and ListMLE [13] both have the property of order sensitive. Then if the probability space is order preserving, these loss functions will be statistically consistent with the permutation-level 0–1 loss.

18.3.3 Regarding Top-k True Loss

In [12], it is argued that the use of a permutation-level 0–1 loss is not appropriate for ranking in information retrieval. Instead, a top-k true loss can better describe the real applications. The top-k true loss takes a value of 0 if the top-k documents and their orders in the ranked list given by the ranking function are exactly the same as the top-k documents in the ground-truth label; and takes a value of 1 otherwise.

By extending the concepts of order preserving and order sensitive in [13], the authors proposed the new concepts of top-k order preserving and top-k order sensitive. The *top-k order preserving property* indicates that if the top-k subgroup probability for a permutation $\pi \in \Omega_{i,j}$ (here $\Omega_{i,j} \triangleq \{\pi \in \Omega : \pi^{-1}(i) < \pi^{-1}(j)\}$) is larger than that for permutation $\sigma_{i,j}^{-1}\pi$, then the relation holds for any other permutation π' in $\Omega_{i,j}$ and the corresponding $\sigma_{i,j}^{-1}\pi'$ provided that the top-k subgroup of the former is different from that of the latter. The *top-k order sensitive* property of a surrogate loss function $L_\phi(f; \mathbf{x}, \pi_y)$ indicates that (i) $L_\phi(f; \mathbf{x}, \pi_y)$ exhibits a symmetry in the sense that simultaneously exchanging the positions of documents i and j in the ground truth and their scores given by the ranking function will not make the surrogate loss change; (ii) when a permutation is transformed to another permutation by exchanging the positions of two documents of it, if the two permutations do not belong to the same top-k subgroup, the loss on the permutation that ranks the two documents in the decreasing order of their scores will not be greater than the loss on its counterpart; (iii) there exists a permutation, for which the speed of change in loss with respect to the score of a document will become faster if exchanging its position with another document with the same score but ranked lower.

Based on the two new concepts, a theorem similar to Theorem 18.3 has been obtained, which is now shown.

Theorem 18.4 *Let $L_\phi(f; \mathbf{x}, \pi_y)$ be a top-k order sensitive loss function. For $\forall m$ documents, if its top-k subgroup probability space is order preserving with respect to $m - 1$ document pairs $\{(j_i, j_{i+1})\}_{i=1}^{k}$ and $\{(j_{k+s_i}, j_{k+i} : 0 \leq s_i < i)\}_{i=2}^{m-k}$, then the loss $L_\phi(f; \mathbf{x}, \pi_y)$ is consistent with the top-k true loss.*

In addition to the above result, the change of statistical consistency with respect to different k values has also been discussed in [12], and the following conclusions are obtained.

- For the consistency with the top-k true loss, when k becomes smaller, the requirement on the probability space becomes weaker but the requirement on the surrogate loss function becomes stronger.
- If we fix the true loss to be top-k and the probability space to be top-k order preserving, the surrogate loss function should be at most top-l ($l \leq k$) order sensitive in order to meet the consistency conditions.

According to Theorem 18.4, the surrogate loss functions optimized by ListMLE and ListNet are not consistent with the top-k true loss. Modifications to the algorithms are needed to make them consistent. For this purpose, one should replace the permutation-level Plackett-Luce model in ListMLE with the top-k Plackett-Luce model; and change the mapping function in ListNet to retain the order for the top-k positions in the ground-truth permutation and assign to all the remaining positions a small value (which is smaller than the score of any document ranked at the top-k positions). Experimental results have shown that with such modifications, the performances of the algorithms can be enhanced in terms of the top-k true loss.

18.3.4 Regarding Weighted Kendall's τ

In [10], the consistency of some pairwise ranking methods and listwise ranking methods with respect to the weighted Kendall's τ is studied. The weighted Kendall's τ is defined as follows:

$$L_0(f; \mathbf{x}, \pi_y) = \sum_{u=1}^{m-1} \sum_{v=u+1}^{m} D(u, v) I_{\{f(x_{\pi_y^{-1}(u)}) - f(x_{\pi_y^{-1}(v)}) \leq 0\}}, \qquad (18.7)$$

where the ground truth is assumed to be a permutation π_y, $D(u, v)$ denotes a weight function on a pair of documents whose positions are u and v in the ground-truth permutation respectively. When $D(u, v) = D(u) - D(v)$, we call it the Difference-Weighting Kendall's τ. When $D(u, v) = 1$, we call it the unweighted Kendall's τ.

Actually the weighted Kendall's τ defined as above has a strong connection with many other evaluation measures or loss functions for ranking. For example, the weighted Kendall's τ is a natural extension of the Kendall's τ, a widely used ranking measure in statistics. The weighted Kendall's τ is also equivalent to DCG with certain gain and discount functions.

Given the weighted Kendall's τ as the true loss for ranking, the existence and uniqueness of the optimal ranking is discussed in [10]. The sufficient conditions for the existence and uniqueness of the optimal ranking are called *separability on pairs* and *transitivity over pairs*. An intuitive explanation of the separability on pairs is that for any document pair all the optimal rankers obtained in learning can separate (rank) the two documents with exactly the same order. The transitivity over pairs guarantees that for any x_s, x_t, x_l, if (x_s, x_t) and (x_t, x_l) are separable with the same order, then (x_s, x_l) is also separable with that order.

With the above two conditions, the statistical consistency for the so-called generalized pairwise surrogate loss function and generalized listwise surrogate loss function are discussed. In particular, the generalized pairwise surrogate loss function is defined by

$$L_\phi(f; \mathbf{x}, \pi_y) = \sum_{u=1}^{m-1} \sum_{v=u+1}^{m} D(u, v) \phi\big(f(x_{\pi_y^{-1}(u)}) - f(x_{\pi_y^{-1}(v)})\big),$$

and the generalization listwise surrogate loss function is defined by

$$L_\phi(f; \mathbf{x}, \pi_y) = \sum_{u=1}^{m-1} D(u) \left[\alpha\big(f(x_{\pi_y^{-1}(u)})\big) + \beta \left(\sum_{v=u}^{m} \gamma\big(f(x_{\pi_y^{-1}(v)})\big) \right) \right],$$

where ϕ is a convex function, $\alpha(\cdot)$ and $\beta(\cdot)$ are non-increasing functions, $\gamma(\cdot)$ is a non-decreasing function.

These two kinds of generalized loss functions can cover the loss functions in Ranking SVM [8, 9], RankBoost [7], RankNet [3] (when $D(u, v) = 1$), and ListMLE [13] ($\alpha(\cdot) = -(\cdot), \beta(\cdot) = -\log(\cdot), \gamma(\cdot) = \exp(\cdot)$) as special cases.

The sufficient conditions for the consistency of these surrogate loss functions are presented in the following two theorems.

Theorem 18.5 *Let $L_\phi(\cdot)$ be a differentiable, non-negative, and non-increasing function with $\phi'(0) < 0$. Then the generalized pairwise surrogate loss function is consistent with the weighted Kendall's τ when $D(u, v)$ are not identical, and is consistent with the unweighted Kendall's τ when $\forall i, j, D(u, v) = c$, where c is a constant.*

Theorem 18.6 *Let $\alpha(\cdot)$ and $\beta(\cdot)$ be non-increasing functions, $\gamma(\cdot)$ be a non-decreasing function. If $\alpha'(x) < 0, \forall x, \beta(\cdot), \gamma(\cdot)$ are differentiable, the generalized listwise surrogate loss function is consistent with the Difference-Weighting Kendall's τ; if $\alpha(\cdot)$ is differentiable, $\forall x, \beta'(x) > 0, \gamma'(x) > 0$, and $\forall i, D(u) = c$, the generalized listwise surrogate loss is consistent with unweighted Kendall's τ.*

Based on the above theorems, we note the following.

- Both RankBoost and RankNet are consistent with the unweighted Kendall's τ (and therefore inconsistent with the weighted Kendall's τ). In order to make them consistent with the weighted Kendall's τ, we need to introduce the position discount factor $D(u, v)$ to each document pair in their surrogate loss functions.
- For Ranking SVM, since the hinge loss does not satisfy the condition in Theorem 18.5, we cannot obtain its consistency based on the current theoretical result.
- ListMLE is consistent with the unweighted Kendall's τ (and therefore inconsistent with the weighted Kendall's τ). If we add the position discount factor $D(u)$ to the likelihood loss, then ListMLE will become consistent with the Difference-Weighting Kendall's τ.

18.4 Consistency Analysis for Two-Layer Ranking

As far as we know, there is no work on the consistency analysis for two-layer ranking.

18.5 Summary

In this chapter, we have introduced some representative work on statistical consistency for ranking. To summarize, we have Table 18.1, which takes the same format as the summary tables in the previous chapter.

According to the table and the content of this chapter, we have the following discussions.

- Given that there is still no consensus on the true loss for ranking, different algorithms may be proven to be consistent with different true losses. In this case, it is still very difficult to directly compare whether the theoretical property of an algorithm is better than that of the other.
- Meaningful true loss should be determined by application. For example, widely used evaluation measures in information retrieval include MAP and NDCG (or DCG). The discussions regarding these measures could be more meaningful than those regarding less frequently used true losses (e.g., pairwise 0–1 loss and permutation-level 0–1 loss). In [6], some discussions are made with respect to

Table 18.1 Consistency analysis for ranking

Approaches	Document ranking	Subset ranking	Two-layer ranking
Pointwise	–	[6]	–
Pairwise	[5]	[10]	–
Listwise	–	[10, 12, 13]	–

DCG, but only the consistency of the regression loss is revealed. It is unknown whether the conclusion can be extended to pairwise and listwise loss functions, which are more popular than the pointwise regression loss.

- Existing analyses are conducted with the setting of either document ranking or subset ranking. However, as pointed out in the previous chapter, the setting of two-layer ranking would make more sense. Unfortunately, there is still no work on the consistency analysis for this setting.

According to the above discussions, there is still a long way to go along the direction of statistical consistency for ranking.

Remark Note that there is some other work highly related to statistical consistency of ranking, although not directly on the topic. For example, in [2], the regret bound between the Kendall's τ and the pairwise 0–1 loss is derived when the ranking result is produced by the voting of the pairwise classification results. In [1], the results in [2] are extended to the case where quick sort is used to produce the ranking result. In [11], the above findings are further extended to the case where NDCG is used as the true loss for ranking instead of the Kendall's τ. Given these theoretical results, if we further consider the previous findings [14] on the consistency of widely used surrogate loss functions like the hinge, exponential, and logistic losses with respect to the 0–1 classification loss, we may also come to the conclusion that in certain conditions the pairwise ranking algorithms like Ranking SVM, RankBoost, and RankNet (which minimize the hinge, exponential, and logistic losses respectively) are consistent with the pairwise 0–1 loss, and thus the Kendall's τ or $(1 - \text{NDCG})$.

18.6 Exercises

18.1 List different true loss functions used in previous work, and compare their pros and cons as the true loss for ranking.

18.2 Can you list the properties that the true loss for ranking should possess, and design such a true loss?

18.3 The definitions of most ranking measures have not considered the sampling of documents. As a result, the extension of them to the setting of two-layer ranking might be difficult. Please show how to extend ranking measures to consider the sampling of documents.

References

1. Ailon, N., Mohri, M.: An efficient reduction from ranking to classification. In: Proceedings of the 21st Annual Conference on Learning Theory (COLT 2008), pp. 87–98 (2008)
2. Balcan, M.F., Bansal, N., Beygelzimer, A., Coppersmith, D., Langford, J., Sorkin, G.B.: Robust reductions from ranking to classification. In: Proceedings of the 20th Annual Conference on Learning Theory (COLT 2007), pp. 139–153 (2007)

3. Burges, C.J., Shaked, T., Renshaw, E., Lazier, A., Deeds, M., Hamilton, N., Hullender, G.: Learning to rank using gradient descent. In: Proceedings of the 22nd International Conference on Machine Learning (ICML 2005), pp. 89–96 (2005)
4. Cao, Z., Qin, T., Liu, T.Y., Tsai, M.F., Li, H.: Learning to rank: from pairwise approach to listwise approach. In: Proceedings of the 24th International Conference on Machine Learning (ICML 2007), pp. 129–136 (2007)
5. Clemencon, S., Lugosi, G., Vayatis, N.: Ranking and empirical minimization of u-statistics. The Annals of Statistics **36**(2), 844–874 (2008)
6. Cossock, D., Zhang, T.: Subset ranking using regression. In: Proceedings of the 19th Annual Conference on Learning Theory (COLT 2006), pp. 605–619 (2006)
7. Freund, Y., Iyer, R., Schapire, R., Singer, Y.: An efficient boosting algorithm for combining preferences. Journal of Machine Learning Research **4**, 933–969 (2003)
8. Herbrich, R., Obermayer, K., Graepel, T.: Large margin rank boundaries for ordinal regression. In: Advances in Large Margin Classifiers, pp. 115–132 (2000)
9. Joachims, T.: Optimizing search engines using clickthrough data. In: Proceedings of the 8th ACM SIGKDD International Conference on Knowledge Discovery and Data Mining (KDD 2002), pp. 133–142 (2002)
10. Lan, Y., Liu, T.Y., Ma, Z.M., Li, H.: Statistical consistency of ranking methods. Tech. rep., Microsoft Research (2010)
11. Sun, Z., Qin, T., Tao, Q., Wang, J.: Robust sparse rank learning for non-smooth ranking measures. In: Proceedings of the 32nd Annual International ACM SIGIR Conference on Research and Development in Information Retrieval (SIGIR 2009), pp. 259–266 (2009)
12. Xia, F., Liu, T.Y., Li, H.: Statistical consistency of top-k ranking. In: Advances in Neural Information Processing Systems 22 (NIPS 2009), pp. 2098–2106 (2010)
13. Xia, F., Liu, T.Y., Wang, J., Zhang, W., Li, H.: Listwise approach to learning to rank—theorem and algorithm. In: Proceedings of the 25th International Conference on Machine Learning (ICML 2008), pp. 1192–1199 (2008)
14. Zhang, T.: Statistical behavior and consistency of classification methods based on convex risk minimization. Annual Statistics **32**(1), 56–85 (2004)

Part VII
Summary and Outlook

In this part, we will give a summary of the book and mention several research directions for future study. We try to conclude the book by answering several important questions regarding existing learning-to-rank algorithms, their empirical performances, and theoretical properties. We present future directions by giving several examples.

After reading this part (as well as the entire book), the readers are expected to have quite a comprehensive view on the topic of learning to rank for information retrieval, to be able to use existing algorithms to solve their real ranking problems, and to be able to invent better learning-to-rank technologies.

Part VII
Summary and Outlook

Chapter 19
Summary

Abstract In this chapter, we summarize the entire book. In particular, we show the example algorithms introduced in this book in a figure. We then provide the answers to several important questions regarding learning to rank raised at the beginning of the book.

In this book, we have given a comprehensive overview of the state of the art in learning to rank.

- We have introduced three major approaches to learning to rank. The first is called the pointwise approach, which reduces ranking to regression, classification, or ordinal regression on each single document. The second is called the pairwise approach, which basically formulates ranking as a classification problem on each document pair. The third is called the listwise approach, which regards ranking as a new problem, and tries to optimize a measure-specific or non-measure-specific loss function defined on all the documents associated with a query. We have introduced the representative algorithms of these approaches, discussed their advantages and disadvantages, and validated their empirical effectiveness on the LETOR benchmark datasets.
- We have mentioned some advanced tasks for learning to rank. These advanced tasks turn out to be more complex than relevance-based document ranking. For example, one needs to consider the diversity in the ranking result, to make use of unlabeled data to improve the training effectiveness, to transfer from one task to another task or from one domain to another domain, and to apply different ranking models to different queries.
- We have discussed the practical issues on learning to rank, such as data processing and applications of learning to rank. Since manual judgment is costly, in many current practices, the ground truths are mined from the click-through logs of search engines. Given that there are noise and position bias in the log data, it is necessary to build some user behavior models to remove their influences. Furthermore, when using an existing learning-to-rank algorithm in real applications, one needs to go through a procedure, including training data construction, feature extraction, query and document selection, feature selection, etc.
- We have introduced the statistical learning theory for ranking, which turns out to be very different from the theories for conventional machine learning problems

like regression and classification. Since the output space of ranking is much more complex than those of regression and classification, and there is a hierarchical structure in the ranking problem (i.e., query-document), the generalization ability and statistical consistency in ranking should be revisited.

As a summary, we plot the representative algorithms of the three approaches introduced in this book in Fig. 19.1. Due to space restrictions, we have not put algorithms on data preprocessing, advanced ranking tasks, and theoretical work on ranking in this figure. From the figure, one can find that learning to rank for information retrieval has become hotter and hotter in recent years, and more and more attention has been paid to the listwise approach.

At the end of this book, let us come back to the questions we raised in the introduction, and provide their possible answers.

(1) *In what respect are these learning-to-rank algorithms similar and in which aspects do they differ? What are the strengths and weaknesses of each algorithm?*

The answer can be found by the algorithm description and categorization. Basically algorithms belonging to the pointwise approach reduce ranking to either regression, classification, or ordinal regression. Algorithms belonging to the pairwise approach reduce ranking to pairwise classification. The advantage of these two approaches is that many existing theories and tools in machine learning can be directly applied. However, distinct properties of ranking have not been considered in their formulations. For example, most evaluation measures for ranking are query level and position based. However, neither the query information nor the positional information is visible to the loss functions of these two approaches. The listwise approach instead treats ranking as a new problem, and defines specific algorithms for it. It can better leverage the concept of *query*, and consider the positional information in the ranking results when training the ranking model. The problem with the listwise approach is that it is in general more complex than the pointwise and pairwise approaches, and a new theoretical foundation is needed.

According to the analysis in Chap. 5, the loss functions of many learning-to-rank methods, no matter whether pointwise, pairwise, or listwise, are upper bounds of $(1 - \text{NDCG})$ and $(1 - \text{MAP})$. Therefore, the minimization of these loss functions can lead to the minimization of $(1 - \text{NDCG})$ and $(1 - \text{MAP})$, or the maximization of NDCG and MAP.

(2) *Empirically speaking, which of those many learning-to-rank algorithms perform the best?*

According to the discussions in Part IV, the LETOR benchmark datasets have been widely used in the recent literature of learning to rank. Due to the standard data collection, feature representation, dataset partitioning, and evaluation tools provided in LETOR, it is possible to perform fair comparisons among different learning-to-rank methods. Empirical studies on LETOR have shown that the listwise ranking algorithms seem to have certain advantages over other algorithms, especially for the top positions of the ranking results, and the pairwise ranking algorithms seem to

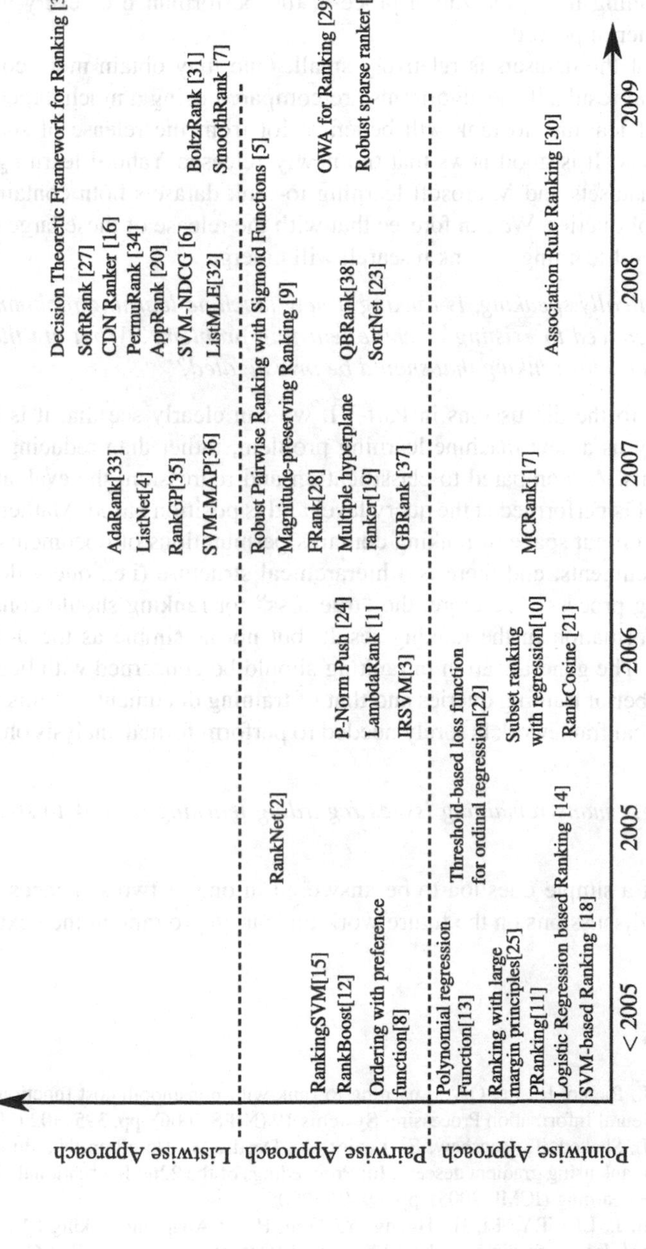

Listwise Approach

- Decision Theoretic Framework for Ranking [39]
- SoftRank [27]
- CDN Ranker [16]
- PermuRank [34]
- AppRank [20]
- SVM-NDCG [6]
- ListMLE[32]
- BoltzRank [31]
- SmoothRank [7]
- AdaRank[33]
- ListNet[4]
- RankGP[35]
- SVM-MAP[36]
- OWA for Ranking [29]
- Robust sparse ranker [26]

Pairwise Approach

- Robust Pairwise Ranking with Sigmoid Functions [5]
- Magnitude-preserving Ranking [9]
- QBRank[38]
- SortNet [23]
- FRank[28]
- Multiple hyperplane ranker[19]
- GBRank [37]
- MCRank[17]
- RankNet[2]
- P-Norm Push [24]
- LambdaRank [1]
- IRSVM[3]
- RankingSVM[15]
- RankBoost[12]
- Ordering with preference function[8]

Pointwise Approach

- Threshold-based loss function for ordinal regression[22]
- Subset ranking with regression[10]
- RankCosine [21]
- Polynomial regression Function[13]
- Ranking with large margin principles[25]
- PRanking[11]
- Logistic Regression based Ranking [14]
- SVM-based Ranking [18]
- Association Rule Ranking [30]

Timeline: <2005   2005   2006   2007   2008   2009

Fig. 19.1 Learning-to-rank algorithms, see [1–39]

outperform the pointwise ranking algorithms. These results are in accordance with the discussions in this book.

Potential problems with the aforementioned experimental results are as follows.

- As pointed out in Chap. 11, these experimental results are still primal and by carefully tuning the optimization process, the performance of every algorithm can be further improved.
- The scale of the datasets is relatively small. One may obtain more convincing experimental results if the algorithms are compared using a much larger dataset. The area of learning to rank will benefit a lot from the release of such large-scale datasets. It is good news that the newly released Yahoo! learning-to-rank challenge datasets and Microsoft learning-to-rank datasets both contain tens of thousands of queries. We can foresee that with the release of these large datasets, a new wave of learning-to-rank research will emerge.

(3) *Theoretically speaking, is ranking a new machine learning problem, or can it be simply reduced to existing machine learning problems? What are the unique theoretical issues for ranking that should be investigated?*

According to the discussions in Part VI, we can clearly see that it is better to regard ranking as a new machine learning problem, rather than reducing it to existing problems. As compared to classification and regression, the evaluation of a ranking model is performed at the query level and is position based. Mathematically speaking, the output space of ranking contains permutations of documents, but not individual documents, and there is a hierarchical structure (i.e., query-document) in the learning process. Therefore, the "true loss" for ranking should consider the positional information in the ranking result, but not as simple as the 0–1 loss in classification. The generalization in ranking should be concerned with both the increasing number of training queries and that of training documents. In this regard, a novel theoretical framework is sorely needed to perform formal analysis on ranking.

(4) *Are there many remaining issues regarding learning to rank to study in the future?*

This is not a simple question to be answered in one or two sentences. We will present more discussions on the future work on learning to rank in the next chapter.

References

1. Burges, C.J., Ragno, R., Le, Q.V.: Learning to rank with nonsmooth cost functions. In: Advances in Neural Information Processing Systems 19 (NIPS 2006), pp. 395–402 (2007)
2. Burges, C.J., Shaked, T., Renshaw, E., Lazier, A., Deeds, M., Hamilton, N., Hullender, G.: Learning to rank using gradient descent. In: Proceedings of the 22nd International Conference on Machine Learning (ICML 2005), pp. 89–96 (2005)
3. Cao, Y., Xu, J., Liu, T.Y., Li, H., Huang, Y., Hon, H.W.: Adapting ranking SVM to document retrieval. In: Proceedings of the 29th Annual International ACM SIGIR Conference on Research and Development in Information Retrieval (SIGIR 2006), pp. 186–193 (2006)

4. Cao, Z., Qin, T., Liu, T.Y., Tsai, M.F., Li, H.: Learning to rank: from pairwise approach to listwise approach. In: Proceedings of the 24th International Conference on Machine Learning (ICML 2007), pp. 129–136 (2007)

5. Carvalho, V.R., Elsas, J.L., Cohen, W.W., Carbonell, J.G.: A meta-learning approach for robust rank learning. In: SIGIR 2008 Workshop on Learning to Rank for Information Retrieval (LR4IR 2008) (2008)

6. Chakrabarti, S., Khanna, R., Sawant, U., Bhattacharyya, C.: Structured learning for nonsmooth ranking losses. In: Proceedings of the 14th ACM SIGKDD International Conference on Knowledge Discovery and Data Mining (KDD 2008), pp. 88–96 (2008)

7. Chapelle, O., Wu, M.: Gradient descent optimization of smoothed information retrieval metrics. Information Retrieval Journal. Special Issue on Learning to Rank 13(3), doi:10.1007/s10791-009-9110-3 (2010)

8. Cohen, W.W., Schapire, R.E., Singer, Y.: Learning to order things. In: Advances in Neural Information Processing Systems 10 (NIPS 1997), vol. 10, pp. 243–270 (1998)

9. Cortes, C., Mohri, M., et al.: Magnitude-preserving ranking algorithms. In: Proceedings of the 24th International Conference on Machine Learning (ICML 2007), pp. 169–176 (2007)

10. Cossock, D., Zhang, T.: Subset ranking using regression. In: Proceedings of the 19th Annual Conference on Learning Theory (COLT 2006), pp. 605–619 (2006)

11. Crammer, K., Singer, Y.: Pranking with ranking. In: Advances in Neural Information Processing Systems 14 (NIPS 2001), pp. 641–647 (2002)

12. Freund, Y., Iyer, R., Schapire, R., Singer, Y.: An efficient boosting algorithm for combining preferences. Journal of Machine Learning Research 4, 933–969 (2003)

13. Fuhr, N.: Optimum polynomial retrieval functions based on the probability ranking principle. ACM Transactions on Information Systems 7(3), 183–204 (1989)

14. Gey, F.C.: Inferring probability of relevance using the method of logistic regression. In: Proceedings of the 17th Annual International ACM SIGIR Conference on Research and Development in Information Retrieval (SIGIR 1994), pp. 222–231 (1994)

15. Herbrich, R., Obermayer, K., Graepel, T.: Large margin rank boundaries for ordinal regression. In: Advances in Large Margin Classifiers, pp. 115–132 (2000)

16. Huang, J., Frey, B.: Structured ranking learning using cumulative distribution networks. In: Advances in Neural Information Processing Systems 21 (NIPS 2008) (2009)

17. Li, P., Burges, C., Wu, Q.: McRank: learning to rank using multiple classification and gradient boosting. In: Advances in Neural Information Processing Systems 20 (NIPS 2007), pp. 845–852 (2008)

18. Nallapati, R.: Discriminative models for information retrieval. In: Proceedings of the 27th Annual International ACM SIGIR Conference on Research and Development in Information Retrieval (SIGIR 2004), pp. 64–71 (2004)

19. Qin, T., Liu, T.Y., Lai, W., Zhang, X.D., Wang, D.S., Li, H.: Ranking with multiple hyperplanes. In: Proceedings of the 30th Annual International ACM SIGIR Conference on Research and Development in Information Retrieval (SIGIR 2007), pp. 279–286 (2007)

20. Qin, T., Liu, T.Y., Li, H.: A general approximation framework for direct optimization of information retrieval measures. Information Retrieval 13(4), 375–397 (2009)

21. Qin, T., Zhang, X.D., Tsai, M.F., Wang, D.S., Liu, T.Y., Li, H.: Query-level loss functions for information retrieval. Information Processing and Management 44(2), 838–855 (2008)

22. Rennie, J.D.M., Srebro, N.: Loss functions for preference levels: regression with discrete ordered labels. In: IJCAI 2005 Multidisciplinary Workshop on Advances in Preference Handling. ACM, New York (2005)

23. Rigutini, L., Papini, T., Maggini, M., Scarselli, F.: Learning to rank by a neural-based sorting algorithm. In: SIGIR 2008 Workshop on Learning to Rank for Information Retrieval (LR4IR 2008) (2008)

24. Rudin, C.: Ranking with a p-norm push. In: Proceedings of the 19th Annual Conference on Learning Theory (COLT 2006), pp. 589–604 (2006)

25. Shashua, A., Levin, A.: Ranking with large margin principles: two approaches. In: Advances in Neural Information Processing Systems 15 (NIPS 2002), pp. 937–944 (2003)

26. Sun, Z., Qin, T., Tao, Q., Wang, J.: Robust sparse rank learning for non-smooth ranking mea-
    sures. In: Proceedings of the 32nd Annual International ACM SIGIR Conference on Research
    and Development in Information Retrieval (SIGIR 2009), pp. 259–266 (2009)
27. Talyor, M., Guiver, J., et al.: Softrank: optimising non-smooth rank metrics. In: Proceedings
    of the 1st International Conference on Web Search and Web Data Mining (WSDM 2008),
    pp. 77–86 (2008)
28. Tsai, M.F., Liu, T.Y., Qin, T., Chen, H.H., Ma, W.Y.: Frank: a ranking method with fidelity
    loss. In: Proceedings of the 30th Annual International ACM SIGIR Conference on Research
    and Development in Information Retrieval (SIGIR 2007), pp. 383–390 (2007)
29. Usunier, N., Buffoni, D., Gallinari, P.: Ranking with ordered weighted pairwise classification.
    In: Proceedings of the 26th International Conference on Machine Learning (ICML 2009),
    pp. 1057–1064 (2009)
30. Veloso, A., Almeida, H.M., Gonçalves, M., Meira, W. Jr.: Learning to rank at query-time using
    association rules. In: Proceedings of the 31st Annual International ACM SIGIR Conference
    on Research and Development in Information Retrieval (SIGIR 2008), pp. 267–274 (2008)
31. Volkovs, M.N., Zemel, R.S.: Boltzrank: learning to maximize expected ranking gain. In: Pro-
    ceedings of the 26th International Conference on Machine Learning (ICML 2009), pp. 1089–
    1096 (2009)
32. Xia, F., Liu, T.Y., Wang, J., Zhang, W., Li, H.: Listwise approach to learning to rank—theorem
    and algorithm. In: Proceedings of the 25th International Conference on Machine Learning
    (ICML 2008), pp. 1192–1199 (2008)
33. Xu, J., Li, H.: Adarank: a boosting algorithm for information retrieval. In: Proceedings of the
    30th Annual International ACM SIGIR Conference on Research and Development in Infor-
    mation Retrieval (SIGIR 2007), pp. 391–398 (2007)
34. Xu, J., Liu, T.Y., Lu, M., Li, H., Ma, W.Y.: Directly optimizing IR evaluation measures in
    learning to rank. In: Proceedings of the 31st Annual International ACM SIGIR Conference on
    Research and Development in Information Retrieval (SIGIR 2008), pp. 107–114 (2008)
35. Yeh, J.Y., Lin, J.Y., et al.: Learning to rank for information retrieval using genetic program-
    ming. In: SIGIR 2007 Workshop on Learning to Rank for Information Retrieval (LR4IR 2007)
    (2007)
36. Yue, Y., Finley, T., Radlinski, F., Joachims, T.: A support vector method for optimizing aver-
    age precision. In: Proceedings of the 30th Annual International ACM SIGIR Conference on
    Research and Development in Information Retrieval (SIGIR 2007), pp. 271–278 (2007)
37. Zheng, Z., Chen, K., Sun, G., Zha, H.: A regression framework for learning ranking functions
    using relative relevance judgments. In: Proceedings of the 30th Annual International ACM
    SIGIR Conference on Research and Development in Information Retrieval (SIGIR 2007),
    pp. 287–294 (2007)
38. Zheng, Z., Zha, H., Zhang, T., Chapelle, O., Chen, K., Sun, G.: A general boosting method
    and its application to learning ranking functions for web search. In: Advances in Neural Infor-
    mation Processing Systems 20 (NIPS 2007), pp. 1697–1704 (2008)
39. Zoeter, O., Taylor, M., Snelson, E., Guiver, J., Craswell, N., Szummer, M.: A decision theo-
    retic framework for ranking using implicit feedback. In: SIGIR 2008 Workshop on Learning
    to Rank for Information Retrieval (LR4IR 2008) (2008)

Chapter 20
Future Work

Abstract In this chapter, we discuss the possible future work on learning to rank. In particular, we show some potential research topics along the following directions: sample selection bias, direct learning from logs, feature engineering, advanced ranking models, large-scale learning to rank, online complexity, robust learning to rank, and online learning to rank. At the end of this chapter, we will make brief discussions on the new scenarios beyond ranking, which seems to be the future trend of search. Algorithmic and theoretical discussions on the new scenario may lead to another promising research direction.

As mentioned several times in the book, there are still many open problems regarding learning to rank. We have made corresponding discussions at the end of several chapters. In addition, there are some other future work items [1], as listed in this chapter. Note that the below list is by no means complete. The field of learning to rank is still growing very fast, and there are a lot more topics awaiting further investigation.

20.1 Sample Selection Bias

Training sets for learning to rank are typically constructed using the so-called *pooling* strategy. These documents are thus, by construction, more relevant than the vast majority of other documents. However, in a search engine, the test process is different. A web search engine typically uses a scheme with two phases (or more) to retrieve the relevant documents. The first phase is a filtering one in which the potentially relevant documents—according to a basic ranking function—are selected from the entire search engine index. Then these documents are scored in a second phase by the learned ranking function. But there is still a large number of documents in this second phase: tens of thousands. And most of these documents have little relevance to the query. There is thus a striking difference in the document distribution between training and test. This problem is called the *sample selection bias* [19]: the documents in the training set have not been drawn at random from the test distribution; they are biased toward relevant documents.

T.-Y. Liu, *Learning to Rank for Information Retrieval*,   241
DOI 10.1007/978-3-642-14267-3_20, © Springer-Verlag Berlin Heidelberg 2011

A standard way of addressing the sample selection bias is to reweight the training samples such that reweighted training distribution matches the test distribution. These weights can be found through logistic regression [2]. Once the weights have been estimated, they can readily be incorporated into a pointwise learning-to-rank algorithm. How to use the weights in a pairwise or listwise algorithm is an interesting research question. Another way of correcting this sample selection bias is to improve the scheme used for collecting training data. The pooling strategy could for instance be modified to include documents deeper in the ranking, thus reflecting more closely the test distribution. But judging more documents has a cost and this brings the question of how to select the training documents under a fixed labeling budget. This has been discussed in Sect. 13.3.

Note that the sample selection bias is related to transfer learning as both deal with the case of different training and test distributions. But in the sample selection bias, even though the marginal distribution $P(x)$ changes between training and test, the conditional output distribution $P(y|x)$ is assumed to be fixed. In most transfer learning scenarios, this conditional output distribution shifts between training and test.

20.2 Direct Learning from Logs

In addition to the sample selection bias, there is another issue with the training sets for learning to rank. That is, the human-labeled training sets are of relatively small scale. Considering the huge query space, even hundreds or thousands of queries might not reasonably guarantee the effectiveness of a learning-to-rank algorithm. Developing very large-scale datasets is very important in this regard.

Click-through log mining is one of the possible approaches to achieve this goal. Several works have been done along this direction, as reviewed in Sect. 13.2. However these works also have certain limitations. Basically they have tried to map the click-through logs to judgments in terms of pairwise preferences or multiple ordered categories. However, this process is not always necessary (and sometimes even not reasonable). Note that there is rich information in the click-through logs, e.g., the user sessions, the frequency of clicking a certain document, the frequency of a certain click pattern, and the diversity in the intentions of different users. After the conversion of the logs to pointwise or pairwise judgments, much of the aforementioned information will be missing. Therefore, it is meaningful to reconsider the problem, and probably change the learning algorithms to adapt to the log data. For example, one can directly regard the click-through logs (without conversion) as the ground truth, and define loss functions based on the likelihood of the log data.

Furthermore, unlike the human-labeled data, click-through logs are of their nature streaming data generated all the time as long as users visit the search engine. Therefore, learning from click-through logs should also be a online process. That is, when the training data shift from human-labeled data to click-through data, the learning scheme should also be changed from offline learning to online learning.

20.3 Feature Engineering

After one extracts a set of features for each document, it seems that the learning-to-rank problem becomes a standard prediction task. However, one should notice that ranking is deeply rooted in information retrieval, so the eventual goal of learning to rank is not only to develop a set of new algorithms and theories, but also to substantially improve the ranking performance in real information retrieval applications. For this purpose, feature engineering cannot be overlooked. It is a killer aspect whether we can encode the knowledge on information retrieval accumulated in the past half a century in the extracted features.

Currently, there is not much work on feature extraction for learning to rank. There may be a couple of reasons. First, the research community of learning to rank has not paid enough attention to this topic, since feature extraction is somehow regarded as engineering while designing a learning algorithm seems to have more research value. However, here we would like to point out that the feature extraction itself also has a lot of research potential. For example, one can study how to automatically extract effective features from raw contents of the query and the web documents. Second, features are usually regarded as the key business secrete for a search engine—it is easy to change for another learning-to-rank algorithm (since different algorithms usually take the same format of inputs), however, it is much more difficult to change the feature extraction pipeline. This is because feature extraction is usually highly coupled with the indexing and storage systems of a search engine, and encodes much human intelligence. As a result, for those benchmark datasets released by commercial search engines, usually we have no access to the features that they really used in their live systems. To work together with these commercial search engines to share more insights on features will be very helpful for the future development of the learning-to-rank community.

20.4 Advanced Ranking Models

In most existing learning-to-rank algorithms, a scoring function is used as the ranking model for the sake of simplicity and efficiency. However, sometimes such a simplification cannot handle complex ranking problems. Researchers have made some attempts on leveraging the inter-relationships between objects [12–14]; however, this is not yet the most straightforward way of defining the hypothesis for ranking, especially for the listwise approach.

Since the output space of the listwise approach is composed of permutations of documents, the ranking hypothesis should better directly output permutations of documents, rather than output scores for each individual document. In this regard, defining the ranking hypothesis as a multi-variate function that directly outputs permutations could be a future research topic. Note that the task is challenging because permutation-based ranking functions can be very complex due to the extremely large number of possible permutations. But it is worthy and also possible to find efficient algorithms to deal with this situation.

The above research direction is related to rank aggregation. In rank aggregation, the input space consists of permutations, and the output space also consists of permutations. There is no feature representation for each single document. The task is to learn the optimal way of aggregating the candidate ranked lists in the input space, to generate the desired ranked list in the output space. There have been several previous works [7, 9] discussing how to use permutation probability models (e.g., the Plackett–Luce model and the Mallows model) to perform unsupervised rank aggregation. It would be an interesting topic to study how to further leverage labeled data and develop supervised versions for these algorithms. One preliminary effort can be found in [11], and we look forward to seeing more such attempts.

Furthermore, most existing work on learning to rank belongs to discriminative learning. However, as we notice, generative learning is also a very important branch of machine learning, which is good at modeling causal relationship or dependency between random variables. There is no reason that generative learning cannot be used in ranking. This could be a promising research direction of learning to rank, from both algorithmic and theoretical points of view.

20.5 Large-Scale Learning to Rank

In the literature of learning to rank, researchers have paid a lot of attention to the design of loss functions, but somehow overlooked the efficiency and scalability of algorithms. The latter, however, has become a more and more important issue nowadays, especially due to the availability of large-scale click-through data in web search that can be used to train the learning-to-rank models.

While it is good to have more training data, it is challenging for many existing algorithms to handle such data. In order to tackle the challenge, we may want to consider one of the following approaches.

- *Parallel computing.* For example, we can use the MPI or MapReduce infrastructure to distribute the computations in the algorithms. There are a number of attempts on distributed machine learning in the literature [4, 5], however, the efforts on learning to rank are still very limited.
- *Ensemble learning.* We can down-sample the data to make it easy to handle by a single-machine algorithm. After learning a ranking model based on this sample, we can repeat the sampling for multiple times and aggregate the ranking models obtained from all the samples. It has been proven in [16] that such ensemble learning can effectively make use of large datasets and the resultant model can be more effective than the model learned from every single sample.
- *Approximate algorithms.* In some cases, we can derive an approximate version of an existing algorithm, whose complexity is much lower than the original algorithm while the accuracy remains to a certain extent. This kind of approaches has been well studied in the literature of computational theory [15], but has still not been sufficiently investigated in learning to rank.

In addition to discussing how to handle large data from a purely computational perspective, we may want to investigate the problem from another angle. That is, it may not be always necessary to increase the data scale. With more and more data, the learning curve may get saturated, and the cost we pay for the extra computations may become wasteful. To gain more understanding on this, we need to jointly consider the learning theory and computational theory for ranking. This can also be an important piece of future work.

20.6 Online Complexity Versus Accuracy

Most of the research on learning to rank has focused on optimizing the relevance of the ranking results. This quest for accuracy can lead to very complex models, and complex models are often computationally expensive. This brings the question on whether this kind of models can be deployed in a real world ranking system where latency is an important factor. In web search for instance there are stringent requirements on the execution time: e.g., the document scoring phase should typically not exceed 50 milliseconds.

For this reason, there is a recent effort on trying to build models which have a reduced execution time. For example, [18] explicitly addressed this accuracy/complexity trade-off for linear ranking functions. A linear function is of course very fast to evaluate, but the feature computation cost should also be taken into account. By performing feature selection, the authors of the aforementioned paper were able to substantially reduce the online complexity without much loss in accuracy.

In the context of decision trees ensembles, execution time can be reduced using *early exits* [3]: if, based on a partial evaluation of the trees, a document does not appear to be relevant, this document is not further evaluated. The ideas in these two papers could be combined using a *cascade* architecture pioneered in [17]. An ensemble of trees would be learned such that early trees use only cheap features while later trees are allowed to use more expensive features. But combined with early exits, these expensive trees would not be evaluated for a large number of documents. This architecture would effectively considerably reduce the online complexity of decision trees ensembles.

This line of research bears some resemblance with large-scale learning. In both cases, the goal is to reduce the time complexity due to large datasets, but large scaling learning is concerned with reducing the *training* time, while the focus of this section is on the *testing* time. Nevertheless, it might be possible to transfer techniques and ideas between these two domains.

20.7 Robust Learning to Rank

In most of the existing works, multiple ranking functions are compared and evaluated, and the best ranking function is selected based on some evaluation measures, e.g. NDCG. For a commercial search engine company, the real scenario is more complex: the ranking function is required to be updated and improved periodically with more training data, newly developed ranking features, or more fancy ranking algorithms, however, the ranking results should not change dramatically. Such requirements would bring new challenge of robustness to learning to rank.

To develop robust ranking model, the very first step is to measure robustness. Multiple evaluations over time is one method but very costly. One practical solution is that robustness can be measured with the probability of switching neighboring pairs in a search result when ranking score turbulence happens [10]. From the perspective of evaluation measure, if adding the robustness factors into the original measures, e.g. NDCG, the new measures could be more suitable to evaluate relevance and robustness at the same time. However, the efforts on robustness measurement are still very preliminary.

How to learn a robust ranking function is another interesting topic. Intuitively, if an algorithm could learn the parameters which control the measure sensitivity to the score turbulence, the generated ranking functions would be more robust. Another possible solution is related to incremental learning, which guarantees that the new model is largely similar to previous models.

20.8 Online Learning to Rank

Traditional learning-to-rank algorithms for web search are trained in a batch mode, to capture stationary relevance of documents to queries, which has limited ability to track dynamic user intention in a timely manner. For those time-sensitive queries, the relevance of documents to a query on breaking news often changes over time, which indicates the batch-learned ranking functions do have limitations. User real-time click feedback could be a better and timely proxy for the varying relevance of documents rather than the editorial judgments provided by human annotators. In other words, an online learning-to-rank algorithm can quickly learn the best re-ranking of the top portion of the original ranked list based on real-time user click feedback.

At the same time, using real-time click feedback would also benefit the recency ranking issue [6], which is to balance relevance and freshness of the top ranking results. Comparing the real-time click feedback over the click history, we could observe some signals that can be leveraged for both ranking and time-sensitive query classification [8].

We have to admit that the research on online learning and recency ranking are still relatively preliminary. How to effectively combine the time-sensitive features into the online learning framework is still an open question for the research community. Furthermore, as a ranking function is frequently updated according to real-time click feedback, how to keep the robustness and stability is another important challenge which cannot be ignored.

20.9 Beyond Ranking

In recent years, there has been a trend in commercial search engines that goes beyond the pure relevance-based ranking of documents in their search results. For example, the computational knowledge engine WolframAlpha.com, the decision engine Bing.com, the universal search in Google.com, all try to provide rich presentation of search result to users.

When the ranked list is no longer the desired output, the learning-to-rank technologies need to be refined: the change of the output space will naturally lead to the change of the hypothesis space and the loss function, as well as the change of the learning theory. On the other hand, the new search scenario may be decomposed into several sub ranking tasks and many key components in learning to rank can still be used. This may become a promising future work for all the researchers currently working on learning to rank, which we would like to call *learning to search* rather than *learning to rank*.

Overall, this book is just a stage-wise summary of the hot research field of learning to rank. Given the fast development of the field, we can foresee that many new algorithms and theories will gradually arrive. We hope that this book will motivate more people to work on learning to rank, so as to make this research direction have more impact in both the information retrieval and machine learning communities.

References

1. Chang, Y., Liu, T.Y.: Future directions in learning to rank. JMLR: Workshop and Conference Proceedings **14**, 91–100 (2011)
2. Bickel, S., Scheffer, T.: Dirichlet-enhanced spam filtering based on biased samples. In: Advances in Neural Information Processing Systems 19 (NIPS 2006) (2007)
3. Cambazoglu, B., Zaragoza, H., Chapelle, O., Chen, J., Liao, C., Zheng, Z., Degenhardt, J.: Early exit optimizations for additive machine learned ranking systems. In: Proceedings of the Third ACM International Conference on Web Search and Data Mining (WSDM 2010), pp. 411–420 (2010)
4. Chang, E., Zhu, K., Wang, H., Bai, H., Li, J., Qiu, Z., Cui, H.: Parallelizing support vector machines on distributed computers. In: Platt, J., Koller, D., Singer, Y., Roweis, S. (eds.) Advances in Neural Information Processing Systems 20 (NIPS 2007), pp. 257–264. MIT Press, Cambridge (2008)
5. Chu, C.T., Kim, S.K., Lin, Y.A., Yu, Y., Bradski, G., Ng, A., Olukotun, K.: Map-reduce for machine learning on multicore. In: Schölkopf, B., Platt, J., Hoffman, T. (eds.) Advances in Neural Information Processing Systems 19 (NIPS 2006), pp. 281–288. MIT Press, Cambridge (2007)
6. Dong, A., Chang, Y., Zheng, Z., Mishne, G., Bai, J., Zhang, R., Buchner, K., Liao, C., Diaz, F.: Towards recency ranking in web search. In: Proceedings of 3rd ACM International Conference on Web Search and Data Mining (WSDM 2010) (2010)
7. Guiver, J., Snelson, E.: Bayesian inference for Plackett–Luce ranking models. In: Proceedings of the 26th International Conference on Machine Learning (ICML 2009), pp. 377–384 (2009)
8. Inagaki, Y., Sadagopan, N., Dupret, G., Liao, C., Dong, A., Chang, Y., Zheng, Z.: Session based click features for recency ranking. In: Proceedings of 24th AAAI Conference on Artificial Intelligence (AAAI 2010) (2010)

9. Klementiev, A., Roth, D., Small, K.: Unsupervised rank aggregation with distance-based models. In: Proceedings of the 25th International Conference on Machine Learning (ICML 2008), pp. 472–479 (2008)
10. Li, X., Li, F., Ji, S., Zheng, Z., Dong, A., Chang, Y.: Incorporating robustness into web ranking evaluation. In: Proceedings of the 18th ACM Conference on Information and Knowledge Management (CIKM 2009) (2009)
11. Qin, T., Geng, X., Liu, T.Y.: A new probabilistic model for rank aggregation. In: Lafferty, J., Williams, C.K.I., Shawe-Taylor, J., Zemel, R., Culotta, A. (eds.) Advances in Neural Information Processing Systems 23 (NIPS 2010), pp. 1948–1956 (2011)
12. Qin, T., Liu, T.Y., Zhang, X.D., Wang, D., Li, H.: Learning to rank relational objects and its application to web search. In: Proceedings of the 17th International Conference on World Wide Web (WWW 2008), pp. 407–416 (2008)
13. Qin, T., Liu, T.Y., Zhang, X.D., Wang, D.S., Li, H.: Global ranking using continuous conditional random fields. In: Advances in Neural Information Processing Systems 21 (NIPS 2008), pp. 1281–1288 (2009)
14. Radlinski, F., Kleinberg, R., Joachims, T.: Learning diverse rankings with multi-armed bandits. In: Proceedings of the 25th International Conference on Machine Learning (ICML 2008), pp. 784–791 (2008)
15. Sipser, M.: Introduction to the Theory of Computation, 2nd edn. PWS Publishing, Boston (2006)
16. Ueda, N., Nakano, R.: Generalization error of ensemble estimators. In: Proceedings of the IEEE International Conference on Neural Networks (ICNN 1996), pp. 90–95 (1996)
17. Viola, P.A., Jones, M.J.: Robust real-time face detection. International Journal of Computer Vision **57**(2), 137–154 (2004)
18. Wang, L., Lin, J., Metzler, D.: Learning to efficiently rank. In: Proceeding of the 33rd International ACM SIGIR Conference on Research and Development in Information Retrieval (SIGIR 2010), pp. 138–145 (2010)
19. Zadrozny, B.: Learning and evaluating classifiers under sample selection bias. In: Proceedings of the Twenty-First International Conference on Machine Learning (ICML 2004) (2004)

Part VIII
Appendix

In this part, we provide some background knowledge for readers to better understand the book. In particular, we first introduce some mathematical tools that are frequently used throughout the book and then introduce basic concepts of supervised machine learning.

Chapter 21
Mathematical Background

In this chapter, we gather some key mathematical concepts for better understanding this book. This is not intended to be an introductory tutorial, and it is assumed that the reader already has some background on probability theory, linear algebra, and optimization.

When writing this chapter, we have referred to [1–4] to a large extent. Note that we will not add explicit citations in the remaining part of this chapter. The readers are highly encouraged to read the aforementioned material.

21.1 Probability Theory

Probability theory plays a key role in machine learning and information retrieval, since the design of learning methods and ranking models often relies on the probability assumption on the data.

21.1.1 Probability Space and Random Variables

When we talk about probability, we often refer to the probability of an uncertain event. Therefore, in order to discuss probability theory formally, we must first clarify what the possible events are to which we would like to attach a probability.

Formally, a probability space is defined by the triple $(\Omega, \mathscr{F}, P)$, where Ω is the space of possible outcomes, $\mathscr{F} \subseteq 2^{\Omega}$ is the space of (measurable) events; and P is the probability measure (or probability distribution) that maps an event $E \in \mathscr{F}$ to a real value between 0 and 1.

Given the outcome space Ω, there are some restrictions on the event space $\mathscr{F}$:

- The trivial event Ω and the empty event $\emptyset$ are all in $\mathscr{F}$;
- The event space $\mathscr{F}$ is closed under (countable) union, i.e., if $E_1 \in \mathscr{F}$ and $E_2 \in \mathscr{F}$, then $E_1 \cup E_2 \in \mathscr{F}$;

T.-Y. Liu, *Learning to Rank for Information Retrieval*,
DOI 10.1007/978-3-642-14267-3_21, © Springer-Verlag Berlin Heidelberg 2011

- The event space $\mathscr{F}$ is closed under complement, i.e., if $E \in \mathscr{F}$, then $\Omega \backslash E \in \mathscr{F}$.

  Given an event space $\mathscr{F}$, the probability measure P must satisfy certain axioms.

- For all $E \in \mathscr{F}$, $P(E) \geq 0$.
- $P(\Omega) = 1$.
- For all $E_1, E_2 \in \mathscr{F}$, if $E_1 \cap E_2 = \emptyset$, $P(E_1 \cup E_2) = P(E_1) + P(E_2)$.

Random variables play an important role in probability theory. Random variables allow us to abstract away from the formal notion of event space, because we can define random variables that capture the appropriate events that we are interested in. More specifically, we can regard the random variable as a function, which maps the event in the outcome space to real values. For example, suppose the event is "it is sunny today". We can use a random variable X to map this event to value 1. And there is a probability associated with this mapping. That is, we can use $P(X = 1)$ to represent the probability that it is really sunny today.

21.1.2 Probability Distributions

Since random variables can take different values (or more specifically map the event to different real values) with different probabilities, we can use a probability distribution to describe it. Usually we also refer to it as the probability distribution of the random variable for simplicity.

21.1.2.1 Discrete Distribution

First, we take the discrete case as an example to introduce some key concepts related to probability distribution.

In the discrete case, the probability distribution specifies the probability for a random variable to take any possible values. It is clear that $\sum_a P(X = a) = 1$.

If we have multiple random variables, we will have the concept of joint distribution, marginal distribution, and conditional distribution. Joint distribution is something like $P(X = a, Y = b)$. Marginal distribution is the probability distribution of a random variable on its own. Its relationship with joint distribution is

$$P(X = a) = \sum_b P(X = a, Y = b).$$

Conditional distribution specifies the distribution of a random variable when the value of another random variable is known (or given). Formally conditional probability of $X = a$ given $Y = b$ is defined as

$$P(X = a | Y = b) = \frac{P(X = a, Y = b)}{P(Y = b)}.$$

With the concept of conditional probability, we can introduce another concept, named independence. Here independence means that the distribution of a random variable does not change when given the value of another random variable. In machine learning, we often make such assumptions. For example, we say that the data samples are independently and identically distributed.

According to the above definition, we can clearly see that if two random variables X and Y are independent, then

$$P(X, Y) = P(X)P(Y).$$

Sometimes, we will also use conditional independence, which means that if we know the value of a random variable, then some other random variables will become independent of each other. That is,

$$P(X, Y|Z) = P(X|Z)P(Y|Z).$$

There are two basic rules in probability theory, which are widely used in various settings. The first is the Chain Rule. It can be represented as follows:

$$P(X_1, X_2, \ldots, X_n) = P(X_1)P(X_2|X_1) \cdots P(X_n|X_1, X_2, \ldots, X_{n-1}).$$

The chain rule provides a way of calculating the joint probability of some random variables, which is especially useful when there is independence across some of the variables.

The second rule is the Bayes Rule, which allows us to compute the conditional probability $P(X|Y)$ from another conditional probability $P(Y|X)$. Intuitively, it actually inverses the cause and result. The Bayes Rule takes the following form:

$$P(X|Y) = \frac{P(Y|X)P(X)}{P(Y)} = \frac{P(Y|X)P(X)}{\sum_a P(Y|X=a)P(X=a)}.$$

21.1.2.2 Continuous Distribution

Now we generalize the above discussions to the continuous case. This time, we need to introduce the concept of the probability density function (PDF). A probability density function, p, is a non-negative, integrable function such that

$$\int p(x)\,dx = 1.$$

The probability of a random variable distributed according to a PDF f is computed as follows:

$$P(a \leq x \leq b) = \int_a^b p(x)\,dx.$$

As for the probability density function, we have similar results to those for probabilities. For example, we have

$$p(y|x) = \frac{p(x, y)}{p(x)}.$$

21.1.2.3 Popular Distributions

There are many well-studied probability distributions. In this subsection, we list a few of them for example.

- Bernoulii distribution: $P(X = x) = p^x(1 - p)^{1-x}$, $x = 0$ or 1.
- Poisson distribution: $P(X = k) = \frac{\exp(-\lambda)\lambda^k}{k!}$.
- Gaussion distribution: $p(x) = \frac{1}{\sqrt{2\pi}\sigma} \exp(-\frac{(x-\mu)^2}{2\sigma^2})$.
- Exponential distribution: $p(x) = \lambda \exp(-\lambda x)$, $x \geq 0$.

21.1.3 Expectations and Variances

Expectations and variances are widely used concepts in probability theory. Expectation is also referred to as mean, expected value, or first moment. The expectation of a random variable, denoted as $E[X]$, is defined by

$$E[X] = \sum_a a P(X = a), \quad \text{for the discrete case;}$$

$$E[X] = \int_{-\infty}^{\infty} x p(x) \, dx, \quad \text{for the continuous case.}$$

When the random variable X is an indicator variable (i.e., it takes a value from $\{0, 1\}$), we have the following result:

$$E[X] = P(X = 1).$$

The expectation of a random variable has the following three properties.

- $E[X_1 + X_2 + \cdots + X_n] = E[X_1] + E[X_2] + \cdots + E[X_n]$.
- $E[XY] = E[X]E[Y]$, if X and Y are independent.
- If f is a convex function, $f(E[X]) \leq E[f(x)]$.

The variance of a distribution is a measure of the spread of it. It is also referred to as the second moment. The variance is defined by

$$\text{Var}(X) = E\big[(X - E[X])^2\big].$$

The variance of a random variable is often denoted as σ^2. And σ is called the standard deviation.

The variance of a random variable has the following properties.

- $\mathrm{Var}(X) = E[X^2] - (E[X])^2$.
- $\mathrm{Var}(aX + b) = a^2 \mathrm{Var}(X)$.
- $\mathrm{Var}(X + Y) = \mathrm{Var}(X)\mathrm{Var}(Y)$, if X and Y are independent.

21.2 Linear Algebra and Matrix Computation

Linear algebra is an important branch of mathematics concerned with the study of vectors, vector spaces (or linear spaces), and functions with vector input and output. Matrix computation is the study of algorithms for performing linear algebra computations, including the study of matrix properties and matrix decompositions. Linear algebra and matrix computation are useful tools in machine learning and information retrieval.

21.2.1 Notations

We start with introducing the following notations:

- $\mathscr{R}^n$: the n-dimensional space of real numbers.
- $A \in \mathscr{R}^{m \times n}$: A is a matrix with m rows and n columns, with real number elements.
- $x \in \mathscr{R}^n$: x is a vector with n elements of real numbers. By default, x denotes a column vector, which is a matrix with n rows and one column. To denote a row vector (or a matrix with one row and n columns), we use the transpose of x, which is written as x^T.
- $x_i \in \mathscr{R}$: the ith element of a vector x.

$$ x = \begin{bmatrix} x_1 \\ x_2 \\ \vdots \\ x_n \end{bmatrix} = \begin{bmatrix} x_1 & x_2 & \cdots & x_n \end{bmatrix}^T . $$

- $a_{ij} \in \mathscr{R}$: the element in the ith row and jth column of a matrix A, which is also written by A_{ij}, $A_{i,j}$, etc.

$$ A = \begin{bmatrix} a_{11} & a_{12} & \cdots & a_{1n} \\ a_{21} & a_{22} & \cdots & a_{2n} \\ \vdots & \vdots & \ddots & \vdots \\ a_{m1} & a_{m2} & \cdots & a_{mn} \end{bmatrix} . $$

- a_j or $A_{:,j}$: the jth column of A.

$$ A = \begin{bmatrix} a_1 & a_2 & \cdots & a_n \end{bmatrix} . $$

- $\bar{a}_i^T$ or $A_{i,:}$: the ith row of A.

$$A = \begin{bmatrix} \bar{a}_1^T \\ \bar{a}_2^T \\ \vdots \\ \bar{a}_m^T \end{bmatrix}.$$

21.2.2 Basic Matrix Operations and Properties

21.2.2.1 Matrix Multiplication

Given two matrices $A \in \mathscr{R}^{m \times n}$ and $B \in \mathscr{R}^{n \times l}$, the matrix multiplication (product) of them is defined by

$$C = AB \in \mathscr{R}^{m \times l},$$

in which

$$C_{ik} = \sum_{j=1}^{n} A_{ij} B_{jk}.$$

Note that the number of columns of A and the number of rows of B should be the same in order to ensure the implementation of matrix multiplication.

The properties of matrix multiplication are listed as below:

- Associative: $(AB)C = A(BC)$.
- Distributive: $A(B + C) = AB + AC$.
- Generally not commutative: $AB \neq BA$.

Given two vectors $x, y \in \mathscr{R}^n$, the *inner product* (or *dot product*) of them is a real number defined as

$$x^T y = \sum_{i=1}^{n} x_i y_i = y^T x \in \mathscr{R}.$$

Given two vectors $x \in \mathscr{R}^m$, $y \in \mathscr{R}^n$, the *outer product* of them is a matrix defined as

$$xy^T = \begin{bmatrix} x_1 y_1 & x_1 y_2 & \cdots & x_1 y_n \\ x_2 y_1 & x_2 y_2 & \cdots & x_2 y_n \\ \vdots & \vdots & \ddots & \vdots \\ x_m y_1 & x_m y_2 & \cdots & x_m y_n \end{bmatrix} \in \mathscr{R}^{m \times n}.$$

Given a matrix $A \in \mathcal{R}^{m \times n}$ and a vector $x \in \mathcal{R}^n$, their product is a vector given by $y = Ax \in \mathcal{R}^m$. The product can be expressed in multiple forms as below.

$$y = \begin{bmatrix} \bar{a}_1^T \\ \bar{a}_2^T \\ \vdots \\ \bar{a}_m^T \end{bmatrix} x = \begin{bmatrix} \bar{a}_1^T x \\ \bar{a}_2^T x \\ \vdots \\ \bar{a}_m^T x \end{bmatrix} = Ax = \begin{bmatrix} a_1 & a_2 & \cdots & a_n \end{bmatrix} \begin{bmatrix} x_1 \\ x_2 \\ \vdots \\ x_n \end{bmatrix}$$

$$= x_1 a_1 + x_2 a_2 + \cdots + x_n a_n.$$

21.2.2.2 Identity Matrix

The *identity matrix* $I \in \mathcal{R}^{n \times n}$ is a square matrix with its diagonal elements equal to 1 and the others equal to 0.

$$I_{ij} = \begin{cases} 1, & i = j \\ 0, & i \neq j \end{cases}$$

Given a matrix $A \in \mathcal{R}^{m \times n}$, we have the following property.

$$AI = A = IA.$$

21.2.2.3 Diagonal Matrix

A *diagonal matrix* $D \in \mathcal{R}^{n \times n}$ is a square matrix with all its non-diagonal elements equal to 0, i.e.,

$$D = \operatorname{diag}(d_1, d_2, \ldots, d_n) \quad \text{or} \quad D_{ij} = \begin{cases} d_i, & i = j, \\ 0, & i \neq j. \end{cases}$$

21.2.2.4 Transpose

Given a matrix $A \in \mathcal{R}^{m \times n}$, the *transpose* of A (denoted by A^T) is obtained by swapping the rows and columns, i.e.,

$$A^T \in \mathcal{R}^{n \times m} \quad \text{where } \left(A^T\right)_{ij} = A_{ji}.$$

The properties of transpose are listed as below:

- $(A^T)^T = A$
- $(A + B)^T = A^T + B^T$
- $(AB)^T = B^T A^T$

21.2.2.5 Symmetric Matrix

If a square matrix $A \in \mathscr{R}^{n \times n}$ holds $A^T = A$, it is called a *symmetric matrix*; if it holds $A^T = -A$, it is called an *anti-symmetric matrix*. Given $A \in \mathscr{R}^{n \times n}$, it is not difficult to verify that $A + A^T$ is symmetric and $A - A^T$ is anti-symmetric. Therefore, any square matrix can be written as the sum of a symmetric matrix and an anti-symmetric matrix, i.e.,

$$A = \frac{1}{2}\left(A + A^T\right) + \frac{1}{2}\left(A - A^T\right).$$

21.2.2.6 Trace

The sum of diagonal elements in a square matrix $A \in \mathscr{R}^{n \times n}$ is called the trace of the matrix, denoted by tr A:

$$\text{tr } A = \sum_{i=1}^{n} A_{ii}.$$

The properties of trace are listed as below:

- Given $A \in \mathscr{R}^{n \times n}$, we have tr $A = $ tr A^T.
- Given $A \in \mathscr{R}^{n \times n}$ and $\alpha \in \mathscr{R}$, we have $\text{tr}(\alpha A) = \alpha \text{ tr } A$.
- Given $A, B \in \mathscr{R}^{n \times n}$, we have $\text{tr}(A + B) = \text{tr } A + \text{tr} B$.
- Given A and B such that AB is a square matrix, we have $\text{tr}(AB) = \text{tr}(BA)$.
- Given matrices $A_1, A_2, \ldots, A_k$ such that $A_1 A_2 \cdots A_k$ is a square matrix, we have
$$\text{tr } A_1 A_2 \cdots A_k = \text{tr } A_2 A_3 \cdots A_k A_1 = \text{tr } A_3 A_4 \cdots A_k A_1 A_2 = \cdots = \text{tr } A_k A_1 \cdots A_{k-1}.$$

21.2.2.7 Norm

A *norm* of a vector is a function $f : \mathscr{R}^n \to \mathscr{R}$ that respects the following four conditions:

- non-negativity: $f(x) \geq 0, \forall x \in \mathscr{R}^n$
- definiteness: $f(x) = 0$ if and only if $x = 0$
- homogeneity: $f(\alpha x) = |\alpha| f(x), \forall x \in \mathscr{R}^n$ and $\alpha \in \mathscr{R}$
- triangle inequality: $f(x) + f(y) \geq f(x + y), \forall x, y \in \mathscr{R}^n$

Some commonly-used norms for vector $x \in \mathscr{R}^n$ are listed as below:

- L_1 norm: $\|x\|_1 = \sum_{i=1}^{n} |x_i|$
- L_2 norm (or Euclidean norm): $\|x\|_2 = \left(\sum_{i=1}^{n} x_i^2\right)^{\frac{1}{2}}$
- L_∞ norm: $\|x\|_\infty = \max_i |x_i|$

The above three norms are members of the L_p norm family, i.e., for $p \in \mathscr{R}$ and $p \geq 1$,

$$\|x\|_p = \left(\sum_{i=1}^{n} |x_i|^p \right)^{\frac{1}{p}}.$$

One can also define norms for matrices. For example, the following norm is the commonly-used Frobenius norm. For a matrix $A \in \mathscr{R}^{m \times n}$,

$$\|A\|_F = \sqrt{\sum_{i=1}^{m} \sum_{j=1}^{n} |A_{ij}|} = \sqrt{\mathrm{tr}(A^T A)}.$$

21.2.2.8 Inverse

A square matrix $A \in \mathscr{R}^{n \times n}$ is called *invertible* or *nonsingular* if there exists a square matrix $B \in \mathscr{R}^{n \times n}$ such that

$$AB = I = BA.$$

In this case, B is uniquely determined by A and is referred to as the *inverse* of A, denoted by A^{-1}. If such a kind of B (or A^{-1}) does not exist, we call A a *non-invertible* or *singular* matrix.

The properties of inverse are listed as below:

- given $A \in \mathscr{R}^{n \times n}$, we have $(A^{-1})^{-1} = A$
- given $A \in \mathscr{R}^{n \times n}$, we have $(A^{-1})^T = (A^T)^{-1} = A^{-T}$
- given $A, B \in \mathscr{R}^{n \times n}$, we have $(AB)^{-1} = B^{-1} A^{-1}$

21.2.2.9 Orthogonal Matrix

Given two vectors $x, y \in \mathscr{R}^n$, they are *orthogonal* if $x^T y = 0$.

A square matrix $A \in \mathscr{R}^{n \times n}$ is an *orthogonal matrix* if its columns are orthogonal unit vectors.

$$A^T A = I = AA^T.$$

The properties of orthogonal matrix are listed as below:

- the inverse of an orthogonal matrix equals to its transpose, i.e., $A^{-1} = A^T$
- $\|Ax\|_2 = \|x\|_2$ for any $x \in \mathscr{R}^n$ and orthogonal $A \in \mathscr{R}^{n \times n}$

21.2.2.10 Determinant

The *determinant* of a square matrix $A \in \mathscr{R}^{n \times n}$ is a function $\det : \mathscr{R}^{n \times n} \to \mathscr{R}^n$, denoted by $\det A$ or $|A|$, i.e.,

$$\det A = \sum_{i=1}^{n} (-1)^{i+j} a_{ij} |A_{\setminus i, \setminus j}| \quad (\forall j = 1, 2, \ldots, n)$$

or

$$\det A = \sum_{j=1}^{n} (-1)^{i+j} a_{ij} |A_{\setminus i, \setminus j}| \quad (\forall i = 1, 2, \ldots, n).$$

Here $A_{\setminus i, \setminus j} \in \mathscr{R}^{(n-1) \times (n-1)}$ is the matrix obtained by deleting the ith row and the jth column from A. Note that the above definition is recursive and thus we need to define $|A| = a_{11}$ for $A \in \mathscr{R}^{1 \times 1}$.

Given $A, B \in \mathscr{R}^{n \times n}$, we have the following properties for the determinant:

- $|A| = |A^T|$
- $|AB| = |A||B|$
- $|A| = 0$ if and only if A is singular
- $|A|^{-1} = \frac{1}{|A|}$ if A is nonsingular

21.2.2.11 Quadratic Form

Given a vector $x \in \mathscr{R}^n$ and a square matrix $A \in \mathscr{R}^{n \times n}$, we call the following scalar a *quadratic form*:

$$x^T A x = \sum_{i=1}^{n} \sum_{j=1}^{n} a_{ij} x_i x_j.$$

As we have

$$x^T A x = \left(x^T A x \right)^T = x^T A^T x = x^T \left(\frac{1}{2} A + \frac{1}{2} A^T \right) x,$$

we may as well regard the matrix in a quadratic form to be symmetric.

Given a vector $x \in \mathscr{R}^n$ and a symmetric matrix $A \in \mathscr{R}^{n \times n}$, we have the following definitions:

- If $x^T A x > 0, \forall x$, A is *positive definite*, denoted by $A \succ 0$.
- If $x^T A x \geq 0, \forall x$, A is *positive semidefinite*, denoted by $A \succeq 0$.
- If $x^T A x < 0, \forall x$, A is *negative definite*, denoted by $A \prec 0$.
- If $x^T A x \leq 0, \forall x$, A is *negative semidefinite*, denoted by $A \preceq 0$.
- Otherwise, A is *indefinite*.

21.2.3 Eigenvalues and Eigenvectors

Given a square matrix $A \in \mathcal{R}^{n \times n}$, a non-zero vector $x \in \mathcal{R}^n$ is defined as an *eigenvector* of the matrix if it satisfies the eigenvalue equation

$$Ax = \lambda x \quad (x \neq 0)$$

for some scalar $\lambda \in \mathcal{R}$. In this situation, the scalar λ is called an *eigenvalue* of A corresponding to the eigenvector x. It is easy to verify that αx is also an eigenvector of A for any $\alpha \in \mathcal{R}$.

To compute the eigenvalues of matrix A, we have to solve the following equation,

$$(\lambda I - A)x = 0 \quad (x \neq 0),$$

which has a non-zero solution if and only if $(\lambda I - A)$ is singular, i.e.,

$$|\lambda I - A| = 0.$$

Theoretically, by solving the above *eigenpolynomial*, we can obtain all the eigenvalues $\lambda_1, \lambda_2, \ldots, \lambda_n$. To further compute the corresponding eigenvectors, we may solve the following linear equation,

$$(\lambda_i I - A)x = 0 \quad (x \neq 0).$$

Given a square matrix $A \in \mathcal{R}^{n \times n}$, its eigenvalues $\lambda_1, \lambda_2, \ldots, \lambda_n$ and the corresponding eigenvectors $x_1, x_2, \ldots, x_n$, we have the following properties:

- $\operatorname{tr} A = \sum_{i=1}^{n} \lambda_i$
- $\det A = \prod_{i=1}^{n} \lambda_i$
- The eigenvalues of a diagonal matrix $D = \operatorname{diag}(d_1, d_2, \ldots, d_n)$ are exactly the diagonal elements $d_1, d_2, \ldots, d_n$
- If A is nonsingular, then the eigenvalues of A^{-1} are $1/\lambda_i$ $(i = 1, 2, \ldots, n)$ with their corresponding eigenvectors x_i $(i = 1, 2, \ldots, n)$.

If we denote

$$X = \begin{bmatrix} x_1 & x_2 & \cdots & x_n \end{bmatrix}, \qquad \Lambda = \operatorname{diag}(\lambda_1, \lambda_2, \ldots, \lambda_n),$$

we will have the matrix form of the eigenvalue equation

$$AX = X\Lambda.$$

If the eigenvectors $x_1, x_2, \ldots, x_n$ are linearly independent, then X is nonsingular and we have

$$A = X \Lambda X^{-1}.$$

In this situation, A is called *diagonalizable*.

21.3 Convex Optimization

When we solve a machine learning problem, it is almost unavoidable that we will use some optimization techniques. In this section, we will introduce some basic concepts and representative algorithms for convex optimization.

21.3.1 Convex Set and Convex Function

A set $C \subseteq \mathscr{R}^n$ is a *convex set* if $\forall x, y \in C$ and $\forall \theta \in [0, 1]$, there holds

$$\theta x + (1 - \theta) y \in C,$$

i.e., the line segments between any pairs of x and y lies in C.

A function $f : \mathscr{R}^n \to \mathscr{R}$ is a *convex function* if the domain of f (denoted by dom f) is a convex set and there holds

$$f\big(\theta x + (1 - \theta) y\big) \leq \theta f(x) + (1 - \theta) f(y)$$

$\forall x, y \in$ dom f and $\theta \in [0, 1]$.

Note that (i) function f is *concave* if $-f$ is convex; (ii) function f is *strictly convex* if dom f is convex and there holds $f(\theta x + (1 - \theta) y) < \theta f(x) + (1 - \theta) f(y)$, $\forall x, y \in$ dom f, $x \neq y$, and $\theta \in [0, 1]$; (iii) from the definition of a convex function one can conclude that *any local minimum is a global minimum for convex functions*.

Here are some examples of convex functions:

- affine: $ax + b$ on $x \in \mathscr{R}$, $\forall a, b \in \mathscr{R}$
- exponential: e^{ax} on $x \in \mathscr{R}$, $\forall a \in \mathscr{R}$
- powers: x^α on $x \in (0, +\infty)$, for $\alpha \in (-\infty, 0] \cup [1, +\infty]$
- powers of absolute value: $|x|^p$ on $x \in \mathscr{R}$, $\forall p \in [1, +\infty]$
- negative logarithm: $-\log x$ on $x \in (0, +\infty)$
- negative entropy: $x \log x$ on $x \in (0, +\infty)$
- affine function: $a^T x + b$ on $x \in \mathscr{R}^n$, $\forall a, b \in \mathscr{R}^n$
- norms: $\|x\|_p = (\sum_{i=1}^n |x_i|^p)^{1/p}$ on $x \in \mathscr{R}^n$ for $p \in [1, +\infty]$; $\|x\|_\infty = \max_i |x_i|$ on $x \in \mathscr{R}^n$
- quadratic: $x^2 + bx + c$ on $x \in \mathscr{R}$, $\forall b, c \in \mathscr{R}$

Here are some examples of *concave* functions:

- affine: $ax + b$ on $x \in \mathscr{R}$, $\forall a, b \in \mathscr{R}$
- powers: x^α on $x \in (0, +\infty)$, for $\alpha \in [0, 1]$
- logarithm: $\log x$ on $x \in (0, +\infty)$

21.3.2 Conditions for Convexity

21.3.2.1 First-Order Condition

Function f is *differentiable* if dom f is open and the gradient $\nabla f(x) = (\frac{\partial f(x)}{\partial x_1}, \frac{\partial f(x)}{\partial x_2}, \dots, \frac{\partial f(x)}{\partial x_n})$ exists $\forall x \in$ dom f. Differentiable f is convex if and only if dom f is convex and there holds

$$f(y) \geq f(x) + \nabla f(x)^T (y - x)$$

$\forall x, y \in$ dom f. This is the *first-order condition for convexity*.

21.3.2.2 Second-Order Condition

Function f is *twice differentiable* if dom f is open and the Hessian $\nabla^2 f(x)$ (which is an n-by-n symmetric matrix with $\nabla^2 f(x)_{ij} = \frac{\partial^2 f(x)}{\partial x_i \partial x_j}, i, j = 1, 2, \dots, n$) exists $\forall x \in$ dom f. Twice differentiable f is convex if and only if dom f is convex and Hessian $\nabla^2 f(x)$ is positive semidefinite (i.e., $\nabla^2 f(x) \succeq 0$), $\forall x \in$ dom f. This is the *second-order condition for convexity*.

21.3.3 Convex Optimization Problem

An optimization problem is convex if the objective function is convex and the feasible set is also convex. That is, a *convex optimization problem* is written as

$$\min_{x \in \mathscr{R}^n} f(x),$$

$$\text{s.t.} \quad g_i(x) \leq 0, \quad i = 1, \dots, m$$

$$h_i(x) = 0, \quad i = 1, \dots, l.$$

where f and $g_i, i = 1, \dots, m$ are convex.

Note that any local optimum of a convex optimization problem is globally optimal. Here are some examples of convex optimization problem.

Linear Programming (LP) A linear programming is a convex optimization problem with affine objective and constraint functions. Its feasible set is a polyhedron.

$$\min_{x \in \mathscr{R}^n} \alpha^T x + \beta,$$

$$\text{s.t.} \quad g_i(x) \leq h_i, \quad i = 1, \dots, m,$$

$$a_i^T x = b_i, \quad i = 1, \dots, l,$$

Quadratic Programming (QP) A quadratic programming is a convex optimization problem with a convex quadratic function as its objective function and with affine constraint functions.

$$\min_{x \in \mathscr{R}^n} \frac{1}{2} x^T Q x + p^T x + r,$$

$$\text{s.t.} \quad g_i(x) \le h_i, \quad i = 1, \ldots, m,$$

$$a_i^T x = b_i, \quad i = 1, \ldots, l,$$

where Q is a symmetric and positive semidefinite matrix.

Semidefinite Programming (SDP) A semidefinite programming is a convex optimization problem with a linear objective function optimized over the intersection of the cone of a group of positive semidefinite matrices with an affine space.

$$\min_{x \in \mathscr{R}^n} \alpha^T x,$$

$$\text{s.t.} \quad x_1 K_1 + \cdots + x_n K_n + G \preceq 0,$$

$$a_i^T x = b_i, \quad i = 1, \ldots, l$$

where G and $K_i, i = 1, \ldots, n$ are the cone of positive semidefinite matrices.

21.3.4 Lagrangian Duality

An optimization problem can be converted to its dual form, called the dual problem of the original optimization problem. Sometimes, it is much easier to solve the dual problem.

Suppose the original optimization problem is written as

$$\min_x f(x),$$

$$\text{s.t.} \quad g_i(x) \le 0, \quad i = 1, \ldots, m,$$

$$h_i(x) = 0, \quad i = 1, \ldots, l.$$

The main idea of *Lagrangian Duality* is to take the constraints in the above optimization problem into account by revising the objective function with a weighted sum of the constraint functions. The Lagrangian associated with the above optimization problem is defined as

$$L(x, \theta, \phi) = f(x) + \sum_{i=1}^{m} \theta_i g_i(x) + \sum_{i=1}^{l} \phi_i h_i(x),$$

where θ_i $(i = 1, \ldots, m)$ and ϕ_i $(i = 1, \ldots, l)$ are Lagrangian multipliers for the constraint inequalities and constraint equations.

The *Lagrangian Dual Function* f_d is defined as the minimum value of the Lagrangian over x,

$$f_d(\theta, \phi) = \min_x L(x, \theta, \phi).$$

Suppose the optimal value of the original optimization problem is f^*, obtained at x^*, i.e., $f^* = f(x^*)$. Then $\forall \theta \geq 0$, considering $g_i(x) \leq 0$ and $h_i(x) = 0$, we have,

$$f_d(\theta, \phi) = \min_x \left\{ f(x) + \sum_{i=1}^{m} \theta_i g_i(x) + \sum_{i=1}^{l} \phi_i h_i(x) \right\} \leq \min_x f(x) = f^*.$$

Therefore, $f_d(\theta, \phi)$ is a lower bound for the optimal value f^* of the original problem. To find the best (tightest) bound, we can solve the following *Lagrangian Dual Problem* to approach f^*,

$$\max_{\theta, \phi} f_d(\theta, \phi),$$

$$\text{s.t.} \quad \theta \geq 0.$$

It is not difficult to see that the dual problem is equivalent to the original problem.

Suppose the optimal value of the dual problem is f_d^*, obtained at (θ^*, ϕ^*), i.e., $f_d^* = f_d(\theta^*, \phi^*)$. Generally, $f_d^* \leq f^*$ since $f_d(\theta, \phi) \leq f^*$. This is called as *weak duality*. If $f_d^* = f^*$ holds, then we claim that *strong duality* holds, indicating that the best bound obtained from the Lagrange dual function is tight.

It can be proven that strong duality holds if the optimization problem respects *Slater's Condition*. Slater's Condition requires that there exists a point $x \in \text{dom} \, f$ such that x is strictly feasible, i.e., $g_i(x) < 0$, $\forall i = 1, \ldots, m$ and $h_i(x) = 0$, $\forall i = 1, \ldots, l$.

21.3.5 KKT Conditions

If a convex optimization problem respects Slater's Condition, we can obtain an optimal tuple (x^*, θ^*, ϕ^*) using the Karush–Kuhn–Tucker (KKT) conditions below. Then x^* is an optimal point for the original problem and (θ^*, ϕ^*) is an optimal point for the dual problem.

$$\begin{cases} g_i(x^*) \leq 0, & i = 1, \ldots, m, \\ h_i(x^*) = 0, & i = 1, \ldots, l, \\ \theta_i^* \geq 0, & i = 1, \ldots, m, \\ \theta_i^* g_i(x^*) = 0, & i = 1, \ldots, m, \\ \nabla f(x^*) + \sum_{i=1}^{m} \theta_i^* \nabla g_i(x^*) + \sum_{i=1}^{l} \phi_i^* \nabla h_i(x^*) = 0. \end{cases}$$

References

1. Bishop, C.M.: Pattern Recognition and Machine Learning. Springer, Berlin (2006)
2. Boyd, S., Vendenberghe, L.: Convex Optimization. Cambridge University Press, Cambridge (2003)
3. Golub, G.H., Loan, C.F.V.: Matrix Computations, 3rd edn. Johns Hopkins University Press, Baltimore (1996)
4. Ng, A.: Lecture notes for machine learning. Stanford University cs229 (2010). http://see. stanford.edu/see/courseinfo.aspx?coll=348ca38a-3a6d-4052-937d-cb017338d7b1s

Chapter 22
Machine Learning

In this chapter, we gather some key machine learning concepts that are related to this book. This is not intended to be an introductory tutorial, and it is assumed that the reader already has some background on machine learning. We will first review some basic supervised learning problems, such as regression and classification, and then show how to use statistical learning theory to analyze their theoretical properties.

When writing this chapter, we have referred to [1–5] to a large extent. Note that we will not add explicit citations in the remaining part of this chapter. The readers are highly encouraged to read the aforementioned materials since this chapter is just a quick review of them.

In general, we use x_i to denote the input variables, usually represented by features, and y_i to denote the output or target variables that we are going to predict. A pair (x_i, y_i) is called a training example, and the set of n training examples $\{(x_i, y_i); i = 1, \ldots, n\}$ is called a training set. We use $\mathcal{X}$ to denote the space of input variables, and $\mathcal{Y}$ the space of output values.

In supervised learning, given a training set, the task is to learn a function $h : \mathcal{X} \mapsto \mathcal{Y}$ such that $h(x)$ is a good predictor for the corresponding value of y. The function h is called a hypothesis.

When the target variable that we are going to predict is continuous, the learning problem is called a regression problem. When y can take on only a small number of discrete values (such as 0 or 1), it is called a classification problem.

22.1 Regression

22.1.1 Linear Regression

Here we take linear regression as an example to illustrate the regression problem. In linear regression, the hypothesis takes the following linear form:

$$h(x) = w^T x.$$

T.-Y. Liu, *Learning to Rank for Information Retrieval*,
DOI 10.1007/978-3-642-14267-3_22, © Springer-Verlag Berlin Heidelberg 2011

Given the training set, in order to learn the hypothesis, a loss function is needed. In particular, the loss function can be defined as follows.

$$L(w) = \frac{1}{2} \sum_{i=1}^{n} \left(w^T x_i - y_i \right)^2.$$

In order to find the optimal w that can minimize the above loss function, one can simply use the gradient descent method, since this loss function is convex. By some simple mathematical tricks, one can get the following update rule for gradient descent, which we usually call the least mean squares (LMS) update rule:

$$w_j = w_j + \alpha \sum_{i=1}^{n} \left(y_i - w^T x_i \right) x_{i,j},$$

where j is the index for the features, and $x_{i,j}$ means the jth feature of input variable x_i.

The above update rule is quite intuitive. The magnitude of the update is proportional to the error term $(y_i - w^T x_i)$. As a result, if we encounter a training example on which the hypothesis can predict the target value in a perfect manner, there will be no need to change the parameters. In contrast, if the prediction made by the hypothesis has a large error, a large change will be made to the parameters.

Note that the above update rule makes use of the entire training set in each step of the update. There is an alternative approach, in which we only use one training example in each step. This is usually called a stochastic gradient descent method, and the original update rule corresponds to the so-called batch gradient descend method. Often the stochastic gradient method gets the loss function close to the minimum much faster than the batch gradient descent method, mainly because it updates the parameter more frequently in the training process. In this regard, especially when the training set is very large, the stochastic gradient descent method is usually preferred. However, when using the stochastic gradient descent method, there is the risk of never converging to the minimum when the parameter w keeps oscillating around the minimum of $L(w)$, although in practice most of the values near the minimum will have been good enough.

22.1.2 Probabilistic Explanation

There are many different ways of explaining the above linear regression algorithm, among which the probabilistic explanation shows that the least-square linear regression is a very natural algorithm.

Let us consider the following relationship between the target variable and the input variable:

$$y_i = w^T x_i + \varepsilon_i,$$

where ε_i is an error term that captures either the unmodeled effects or random noise. Since we usually do not know much about this error term, a simple way is to assume that the ε_i are independently and identically distributed (i.i.d.) according to a Gaussian distribution: $\varepsilon_i \sim N(0, \sigma^2)$. In other words, we assume that the probability density of ε_i is

$$p(\varepsilon_i) = \frac{1}{\sqrt{2\pi}\sigma} \exp\left(-\frac{\varepsilon_i^2}{2\sigma^2}\right).$$

Accordingly, we have

$$p(y_i|x_i; w) = \frac{1}{\sqrt{2\pi}\sigma} \exp\left(-\frac{(y_i - w^T x_i)^2}{2\sigma^2}\right).$$

Given the above assumptions, we can write the conditional likelihood of the training data as

$$l(w) = \prod_{i=1}^{m} p(y_i|x_i; w) = \prod_{i=1}^{m} \frac{1}{\sqrt{2\pi}\sigma} \exp\left(-\frac{(y_i - w^T x_i)^2}{2\sigma^2}\right).$$

The log likelihood can then be written as

$$\log l(w) = n \log \frac{1}{\sqrt{2\pi}\sigma} - \frac{1}{\sigma^2} \cdot \frac{1}{2} \sum_{i=1}^{n}(y_i - w^T x_i)^2.$$

Now we maximize this log likelihood in order to get the optimal parameter w. It is not difficult to see that this is equivalent to minimizing the following least-square loss function:

$$L(w) = \frac{1}{2} \sum_{i=1}^{n}(w^T x_i - y_i)^2.$$

The above analysis shows that under certain probabilistic assumptions on the data, least-square regression corresponds to finding the maximum likelihood estimation of w.

22.2 Classification

The literature of classification is relatively richer than that of regression. Many classification methods have been proposed, with different loss functions and different formulations. In this section, we will take binary classification as an example to illustrate several widely used classification algorithms, which are mostly related to this book.

Fig. 22.1 Neural networks

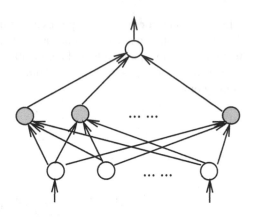

22.2.1 Neural Networks

Neural networks provide a general method for learning real-valued, discrete-valued, and vector-valued functions from examples. Algorithms such as Back Propagation (BP) use gradient descent to tune the network parameters to best fit a training set, and has proven surprisingly successful in many practical problems like speech recognition and handwriting recognition.

In this subsection, we will take multi-layer neural networks as example to illustrate the basic idea of neural networks and the BP algorithm.

A typical multi-layer neural network is shown in Fig. 22.1. The first layer corresponds to the input vectors; the second layer contains hidden nodes which determine the representability of the network; the third layer corresponds to the output. Note that if we define each node in the network as a linear function, the multi-layer structure actually will not bring too much difference to a single-layer network, since it can still provide only linear functions. To avoid this problem, a sigmoid unit is used in the network. The sigmoid unit first computes the linear combination of its inputs, then applies a sigmoid function to the result. Specifically,

$$o = \frac{1}{1 + e^{-w^T x}}.$$

Usually the BP method is used to learn the weight for neural networks, given the fixed set of nodes and edges in the networks. Specifically, the BP algorithm repeatedly iterates over the training examples. For each training example, it applies the networks to the example, calculates the error of the networks' output for this example, computes the gradient with respect to the error on the example, and then updates all the weights in the networks. This gradient descent step is iterated until the networks perform acceptably well.

Note that the advantage of using gradient descent is its simplicity. Especially when the sigmoid unit is used in the neural networks, the computation of the gradient becomes highly feasible. The disadvantage is that the gradient descent can only find a local optimum of the parameters. Usually one needs to perform random restarts

Fig. 22.2 Support vector machines

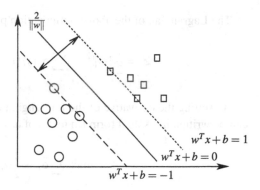

to avoid bad local optimum. However, the time complexity will correspondingly increase.

22.2.2 Support Vector Machines

Intuitively a support vector machine constructs a hyperplane in a high or infinite dimensional space, which can be used for classification. A good separation is achieved by the hyperplane that has the largest distance (i.e., margin) to the nearest training sample of any class, since in general the larger the margin the lower the generalization error of the classifier. In this regard, support vector machines belong to large-margin classifiers.

For ease of discussion, we start with the linear model $w^T x$. If the training data are linearly separable, then the following constraints are to be satisfied:

$$y_i (w^T x_i + b) \geq 1.$$

Given the constraints, the goal is to find the hyperplane with the maximum margin. It is not difficult to see from Fig. 22.2 that the margin can be represented by $\frac{2}{\|w\|}$. Therefore, we have the following optimization problem:

$$\min \frac{1}{2} \|w\|^2,$$

$$\text{s.t.} \quad y_i (w^T x_i + b) \geq 1.$$

If the data are not linearly separable, the constraints cannot be satisfied. In this case, the concept of the soft margin is introduced, and the optimization problem becomes

$$\min \frac{1}{2} \|w\|^2 + C \sum_i \xi_i,$$

$$\text{s.t.} \quad y_i (w^T x_i + b) \geq 1 - \xi_i, \quad \xi_i \geq 0.$$

The Lagrangian of the above optimization problem is

$$\mathcal{L} = \frac{1}{2}\|w\|^2 - \sum_i \alpha_i \left(y_i \left(w^T x_i + b \right) - 1 + \xi_i \right).$$

By setting the derivative of the Lagrangian to be zero, the optimization problem can be written in its dual form (in terms of α_i),

$$\max \sum_i \alpha_i - \frac{1}{2}\alpha_i\alpha_j y_i y_j x_i^T x_j,$$

$$\text{s.t.} \quad \sum_i \alpha_i y_i = 0, \quad \alpha_i \geq 0.$$

This is a typical quadratic programming problem. After solving it, we can recover the ranking model $w = \sum_i \alpha_i y_i x_i$.

If we use the Karush–Kuhn–Tuker (KTT) conditions to analyze the optimality of the solutions to the above optimization problem, we will come to the following conclusion:

$$\alpha_i = 0 \quad \Rightarrow \quad y_i \left(w^T x_i + b \right) \geq 1,$$

$$0 < \alpha_i < C \quad \Rightarrow \quad y_i \left(w^T x_i + b \right) = 1,$$

$$\alpha_i = C \quad \Rightarrow \quad y_i \left(w^T x_i + b \right) \leq 1.$$

That is, the hyperplane w is the linear combination of a small number of samples. These samples x_i with non-zero α_i are called support vectors. This is why we have the name "support vector machines" for the algorithm.

Note that we have explained support vector machines using the linear model as an example. One can replace the inner product $x_i^T x_j$ in the above discussions with a kernel function $K(x_i, x_j)$ to extend the algorithm to the non-linear case. Typical kernel functions include the polynomial kernel and the RBF kernel. Kernel functions can also be regarded as similarity measures between input objects, just as the inner product. The well-known Mercer's condition states that any positive semi-definite kernel $K(\cdot, \cdot)$, i.e., $\sum_{i,j} K(x_i, x_j)c_i c_j \geq 0$, can be expressed as an inner product in a high dimensional space. Therefore, by using the kernel function, we actually significantly increase the dimensionality of the input space, however, at the same time, keep the computational complexity almost unchanged.

Support vector machines have been applied in many applications, and have achieved great success. This is in part because the algorithm is very simple and clear, and in part because the algorithm has its theoretical guarantee on the generalization ability. The generalization ability is related to the concept of the VC dimension, which will be explained more in the next section.

22.2.3 Boosting

Boosting combines a set of weak learners to create a single strong learner. A weak learner is defined as a classifier that is only slightly correlated with the true classification (it can label examples better than random guess). In contrast, a strong learner is a classifier that is arbitrarily well correlated with the true classification.

Here we take AdaBoost as an example to illustrate the basic idea of Boosting algorithms. AdaBoost consists of iteratively learning weak classifiers with respect to a distribution and adding them to a final strong classifier. When a weak learner is added, it is weighted in some way that is related to its accuracy. After a weak learner is added, the data are reweighted: examples that are misclassified will gain weight and examples that are classified correctly will lose weight. Thus, future weak learners focus more on the examples that previous weak learners misclassified.

Specifically, AdaBoost performs the learning in an iterative manner ($t = 1, \ldots, T$). The algorithm maintains a distribution or set of weights over the training set. The weight of this distribution on training example x_i on round t is denoted $\mathcal{D}_t(i)$. Initially, all weights are set equally, but on each round, the weights of incorrectly classified examples are increased. The details of the AdaBoost algorithm are shown in Algorithm 1.

Algorithm 1: Learning algorithm for AdaBoost

Input: training instances $\{(x_i, y_i)\}_{i=1}^{n}$
Given: initial distribution $\mathcal{D}_1(x_i) = \frac{1}{n}$
For $t = 1, \ldots, T$
  Train weak learner f_t based on distribution $\mathcal{D}_t$
  Get weak learner $f_t : \mathscr{X} \mapsto \{-1, 1\}$ based on distribution $\mathcal{D}_t$, whose error
is $\varepsilon_t = \Pr_{\mathcal{D}_t}[f_t(x_i) \neq y_i]$
  Choose $\alpha_t = \frac{1}{2}\log(\frac{1-\varepsilon_t}{\varepsilon_t})$
  Update $\mathcal{D}_{t+1}(x_i) = \frac{1}{Z_t}\mathcal{D}_t(x_i)\exp(-\alpha_t y_i f_t(x_i))$, where Z_t is a normalizer
Output: $f(x) = \sum_t \alpha_t f_t(x)$

Previous work shows that Boosting has state-of-the-art classification performance, and can avoid over fitting in some cases. That is, when more iterations are conducted, even if the training error has been zero, the test error can still be reduced. This phenomenon has been explained using the margin theory. That is, when more iterations are used, although one cannot further improve the training error, the margin of the training data (or the classification confidence on the training data) can still be improved.

22.2.4 K Nearest Neighbor (KNN)

KNN classifier is an instance-based learning algorithm that is based on a distance function for pairs of observations, such as the Euclidean distance and the Manhattan distance. The basic idea of KNN is very simple. In this classification paradigm, k nearest neighbors of a test sample are retrieved first. Then the similarities between the test sample and the k nearest neighbors are aggregated according to the class of the neighbors, and the test sample is assigned to the most similar class.

The best choice of k depends upon the data; generally, larger values of k reduce the effect of noise on the classification, but make boundaries between classes less distinct. A good k can be selected by various heuristic techniques, for example, cross-validation. The special case where the class is predicted to be the class of the closest training sample (i.e. when $k = 1$) is called the nearest neighbor algorithm.

The accuracy of the KNN algorithm can be severely degraded by the presence of noisy or irrelevant features, or if the feature scales are not consistent with their importance. This is mainly because the similarity measure in KNN uses all features equally. Much research effort has been put into selecting or scaling features to improve classification.

The naive version of the KNN algorithm is easy to implement by computing the distances from the test sample to all stored training samples, but it is computationally intensive, especially when the size of the training set grows. Many nearest neighbor search algorithms have been proposed over the years; these generally seek to reduce the number of distance evaluations actually performed. Using an appropriate nearest neighbor search algorithm makes KNN computationally tractable even for large datasets.

The nearest neighbor algorithm has some strong consistency results. As the amount of data approaches infinity, the algorithm is guaranteed to yield an error rate no worse than twice the Bayes error rate (the minimum achievable error rate given the distribution of the data).

22.3 Statistical Learning Theory

Statistical learning theory provides a framework for studying the problems of gaining knowledge, making predictions, making decisions, or constructing models from a set of data. Statistical learning theory is to formalize the above problems, to describe existing learning algorithms more precisely, and to guide the development of new or improved algorithms.

In particular, when we are interested in supervised learning, the task is as follows. Given a set of training data consisting of label-instance pairs (e.g., the label is either $+1$ or -1), a model is automatically constructed from the data using a specific algorithm. The model maps instances to labels, and should make few mistakes when predicting the labels of unseen instances. Of course, it is always possible to build a model that exactly fits the training data. However, in the presence of noise, this might not be the best choice since it would lead to poor performances on the unseen instances. To tackle the problem, many learning algorithms consider the generalization of the model from training to test data. For this purpose, one should choose those models that well fit the data, but at the same time are as simple as possible (to avoid over fitting on the training data). Then an immediate question is how to measure the simplicity (or its counterpart complexity) of a model. This is one of the central problems in statistical learning theory.

Regarding this question, one needs to address an important issue: the assumption on the data. For this issue, one may need to be aware of the famous *No Free Lunch* theory, which basically states that if there is no assumption on the connection between training and test data, prediction is impossible. Furthermore, if there are no priori assumptions on possible test data that are expected, it is impossible to generalize and there is thus no better algorithm (any algorithm will be beaten by some other ones on certain kind of test data). Therefore, one needs to make a reasonable assumption on the data before deriving learning algorithms and performing theoretical analysis on them.

In statistical learning theory, the assumption on the data is that both training and test data are sampled independently from the same distribution. The independence assumption means that all new data yield maximum information, and the identical distribution means that the new data give information about the same underlying probability distribution.

In the rest of the section, we will take binary classification as an example to show some basic concepts of statistical learning theory.

22.3.1 Formalization

In this subsection, we formalize the basic problems in statistical learning theory. For simplicity, we consider binary classification, where the input space is $\mathcal{X}$ and the output space is $\mathcal{Y} = \{+1, -1\}$. We assume that the pairs $(X, Y) \in \mathcal{X} \times \mathcal{Y}$ are random variables distributed according to an unknown distribution P and the goal is to construct a function $g : \mathcal{X} \to \mathcal{Y}$ which predicts Y from X.

For randomly chosen test data, the error made by the function g can be represented as

$$R(g) = P\big(g(X) \neq Y\big) = E[I_{\{g(X) \neq Y\}}].$$

It is clear that if we introduce the regression function $\eta(x) = E[Y|X = x] = 2P(Y = 1|X = x) - 1$ and the target function (or Bayes classifier) $t(x) = \mathrm{sgn}(\eta(x))$, then this target function would achieve the minimum risk over all the possible measurable functions:

$$R(t) = \inf_g R(g).$$

We denote the value of $R(t)$ as R^*, and refer to it as the Bayes risk.

Our goal is to find this target function $t(x)$, however, since P is unknown we cannot know directly the value of t at the data points. We can only use the following empirical risk to estimate t:

$$\hat{R}(g) = \frac{1}{n} \sum_{i=1}^{n} I_{\{g(X_i) \neq Y_i\}}.$$

However, it might not be a good choice to directly minimize the empirical risk in the entire function class due to the risk of over fitting. There are usually two ways of tackling the challenges. The first is to restrict the class of functions in which the minimization is performed, and the second is to modify the criterion to be minimized (e.g., adding a penalty for complex functions). These two ways can also be combined in some algorithms.

Considering the aforementioned ways of avoiding over fitting, researchers usually design their learning algorithms according to the following paradigms.

Empirical Risk Minimization. Algorithms belonging to this category are the most straightforward yet usually very effective. The idea is to restrict the function class to be $\mathcal{G}$ when performing the minimization of the empirical risk. That is,

$$g_n = \arg\min_{g \in \mathcal{G}} \hat{R}(g).$$

Structural Risk Minimization. Algorithms belonging to this category introduce a penalty of the size of the function class. Suppose there is an infinite sequence $\{\mathcal{G}_d : d = 1, 2, \ldots\}$ of function classes with increasing size, and the learning algorithm actually minimizes the following objective function:

$$g_n = \arg\min_{g \in \mathcal{G}_d} \hat{R}(g) + Pen(d, n)$$

where the penalty $Pen(d, n)$ gives preference to the models from a smaller function class.

Sometimes, an alternative and easier-to-implement approach to the structural risk minimization is to adopt a regularization term (typically the norm of g) in the objective function, as follows:

$$g_n = \arg\min_{g \in \mathcal{G}} \hat{R}(g) + \lambda \|g\|^2.$$

As can be seen above, the eventual goal is to minimize the expected risk $R(g)$ while the empirical risk $\hat{R}(g)$ (or structural risk) is actually minimized by learning algorithms. Given the learned function $g_n \in \mathcal{G}$, in order to evaluate whether it is good or not, one still needs to look at $R(g_n)$. However, $R(g_n)$ is a random variable and cannot be computed from the data. Therefore, one can usually only take the form of probabilistic bounds for the analysis on the learning algorithms. In the following paragraphs, we will introduce the basic forms of such bounds.

Here we introduce the best function $g^* \in \mathcal{G}$ with $R(g^*) = \inf_{g \in \mathcal{G}} R(g)$, we have

$$R(g_n) - R^* = \left[R(g^*) - R^* \right] + \left[R(g_n) - R(g^*) \right].$$

The first term on the right-hand side of the above equation is called the approximation error, which measures how well the best function in $\mathscr{G}$ can approach the target. The second term, called the estimation error, measures how close g_n is to the best function in $\mathscr{G}$. In the literature of statistical learning theory, most attention has been paid to the estimation error, which is also the focus of our introduction in this section.

The bound regarding the estimation error takes the form

$$R(g_n) \leq R(g^*) + B(n, \mathscr{G}).$$

Considering that g_n minimizes the empirical risk in $\mathscr{G}$, we have

$$\hat{R}(g^*) - \hat{R}(g_n) \geq 0.$$

Then we obtain the inequality

$$R(g_n) = R(g_n) - R(g^*) + R(g^*)$$
$$\leq \hat{R}(g^*) - \hat{R}(g_n) + R(g_n) - R(g^*) + R(g^*)$$
$$\leq 2 \sup_{g \in \mathscr{G}} \left| R(g) - \hat{R}(g) \right| + R(g^*).$$

Based on the above inequality, it is clear that if one can obtain a bound for $|R(g) - \hat{R}(g)|$, one will correspondingly obtain a bound for the estimation error. This is what many previous works have been doing.

22.3.2 Bounds for $|R(g) - \hat{R}(g)|$

In order to ease the analysis on $|R(g) - \hat{R}(g)|$, here we define a new function class called the loss class:

$$\mathscr{F} = \left\{ f : (x, y) \mapsto I_{\{g(x) \neq y\}}, g \in \mathscr{G} \right\}.$$

Then it is clear that $R(g) = E[f(Z)]$, and $\hat{R}(g) = \frac{1}{n} \sum_{i=1}^{n} f(Z_i)$, where $Z = (X, Y)$ and $Z_i = (X_i, Y_i)$.

22.3.2.1 Hoeffding's Inequality

One may have noticed that the relationship between $R(g)$ and $\hat{R}(g)$ is actually the relationship between expectation and average. According to the law of large number, when the number of samples approaches infinity, $\hat{R}(g)$ will converge to $R(g)$. Furthermore, there is actually a quantitative version of the law of large number, when the function g is bounded. This is the famous Hoeffding's Inequality.

Theorem 22.1 *Let* $Z_1, \ldots, Z_n$ *be* n *i.i.d. random variables with* $f(Z) \in [a, b]$. *Then* $\forall \varepsilon > 0$, *we have*

$$P\left[\left|\frac{1}{n}\sum_{i=1}^{n} f(Z_i) - E[f(Z)]\right| > \varepsilon\right] \leq 2\exp\left(-\frac{2n\varepsilon^2}{(b-a)^2}\right).$$

If we denote the right-hand side of the Hoeffding's Inequality as δ, we can have the following equivalent form of it: with probability at least $1 - \delta$,

$$\left|\frac{1}{n}\sum_{i=1}^{n} f(Z_i) - E[f(Z)]\right| \leq (b-a)\sqrt{\frac{\log\frac{2}{\delta}}{2n}}.$$

In our case, $f(Z) \in [0, 1]$. Further considering the relationship between f and g, we have with probability at least $1 - \delta$,

$$R(g) \leq \hat{R}(g) + \sqrt{\frac{\log\frac{2}{\delta}}{2n}}.$$

The above bound looks quite nice, however, it should be noted that the bound also has its limitations. The major issue lies in that Hoeffding's Inequality holds for a fixed function f, and the probability is with regards to the sampling of the data. That is, given a function f, it is of large probability that there is a set of samples satisfying the inequality. However, these sets can be different for different functions. As a result, for a fixed sample, only some of the functions in $\mathscr{F}$ will satisfy the inequality.

To tackle the aforementioned problem, one choice is to consider the supremum of all the functions and derive a looser bound. We will make more introductions on this in the next subsection.

22.3.2.2 Uniform Bounds in Finite Case

In order to obtain a uniform bound which is not data dependent, we need to consider the failure case for each function in the class. For ease of discussion, here we assume that there are a finite number of functions in the class, and denote the number as N. Then, for f_k $(k = 1, \ldots, N)$, we use C_i to denote the set of "bad" samples, i.e., those for which the bound given by Hoeffding's inequality fails. That is,

$$C_k = \left\{z : \left|\frac{1}{n}\sum_{i=1}^{n} f_k(Z_i) - E[f_k(Z)]\right| > \varepsilon\right\}.$$

It is clear that $P(C_k) \leq \delta$. Then according to the properties of probabilities of union sets, we can obtain that $P(C_1 \cup \cdots \cup C_N) \leq N\delta$. As a result, we have

$$P\left(\exists f \in \{f_1, \ldots, f_N\} : \left|\frac{1}{n}\sum_{i=1}^{n} f_k(Z_i) - E\big[f_k(Z)\big]\right| > \varepsilon\right)$$

$$\leq NP\left(\left|\frac{1}{n}\sum_{i=1}^{n} f_k(Z_i) - E\big[f_k(Z)\big]\right| > \varepsilon\right) \leq N\exp(-2n\varepsilon^2).$$

Equivalently, we have $\forall 0 < \delta < 1$, with probability at least $1 - \delta$,

$$\forall g \in \{g_1, \ldots, g_N\}, R(g) \leq \hat{R}(g) + \sqrt{\frac{\log \frac{2N}{\delta}}{2n}}.$$

22.3.2.3 Uniform Bounds in Infinite Case

While the bound introduced in the previous subsection can successfully handle the finite function class, when the number of functions approaches infinity, it becomes unclear whether the bound can converge. To better bound the difference between empirical and expected risks in the infinite case, one needs to adopt some new concepts, such as the VC dimension.

When the function class is uncountable (when it is countable, a trivial extension of the results presented in the previous subsection can be directly applied and there is no need to introduce the VC dimension), we look at the function class projected on the sample. Specifically, given a sample $z_1, \ldots, z_n$, we consider

$$\mathscr{F}_{z_1,\ldots,z_n} = \big\{(f(z_1), \ldots, f(z_n)) : f \in \mathscr{F}\big\}.$$

The size of this set is always finite, since function f can only take binary values. An intuitive explanation on the size of the set is the number of possible ways in which the data $z_1, \ldots, z_n$ can be classified. Based on the set, we can define the concept of a growth function as follows.

Definition 22.1 The growth function is the maximum number of ways into which n data samples can be classified by the function class:

$$S_{\mathscr{F}}(n) = \sup_{z_1,\ldots,z_n} |\mathscr{F}_{z_1,\ldots,z_n}|.$$

It has been proven that if we define the growth function for $\mathscr{G}$ in the same way, we will have $S_{\mathscr{F}}(n) = S_{\mathscr{G}}(n)$.

Given the concept of the growth function, the following result has been obtained:

Theorem 22.2 $\forall 0 < \delta < 1$, with probability at least $1 - \delta$,

$$\forall g \in \{g_1, \ldots, g_N\}, \quad R(g) \leq \hat{R}(g) + 2\sqrt{2\frac{\log S_{\mathscr{G}}(2n) + \log \frac{2}{\delta}}{n}}.$$

In order to further bound the term $\log S_{\mathscr{G}}(2n)$, the concept of VC dimension has been introduced and discussed for typical function classes.

Definition 22.2 The VC dimension of a class $\mathscr{G}$ is the largest n such that $S_{\mathscr{G}}(n) = 2^n$.

In other words, the VC dimension is the size of the largest set that the function class $\mathscr{G}$ can shatter. With the concept of the VC dimension, the following theorem has been proven, which gives an upper bound for the growth function.

Theorem 22.3 *Let $\mathscr{G}$ be a function class whose VC dimension is h. Then $\forall n$,*

$$S_{\mathscr{G}}(n) \le \sum_{i=0}^{h} C_n^i,$$

and $\forall n \ge h$,

$$S_{\mathscr{G}}(n) \le \left(\frac{en}{h}\right)^h.$$

Based on the above theorem, we can obtain the uniform bound for the infinite case, as shown below. $\forall 0 < \delta < 1$, with probability at least $1 - \delta$,

$$\forall g \in \mathscr{G}, \quad R(g) \le \hat{R}(g) + 2\sqrt{2\frac{h\log\frac{2en}{h} + \log\frac{2}{\delta}}{n}}.$$

In addition to the VC dimension introduced in the previous subsection, there are also several other measures which may not only give tighter bounds, but also have properties that make their computations possible from the data only. Specifically we will introduce the covering number and Rademacher average, as well as the bounds that they can lead to in the next subsections.

22.3.2.4 Uniform Bounds Based on Covering Number

We first define the normalized Hamming distance of the projected functions on the sample as follows:

$$d_n(f, f') = \frac{1}{n}\left|\{f(z_i) \ne f'(z_i) : i = 1, \ldots, n\}\right|.$$

Then given such a distance, we say that a set of functions $f_1, \ldots, f_N$ can cover $\mathscr{F}$ at radius ε if

$$\mathscr{F} \subset B(f_i, \varepsilon)$$

where $B(f_i, \varepsilon)$ is the ball whose center is f_i and radius measured by d_n is ε.

Correspondingly, we define the covering number of $\mathscr{F}$ as follows.

Definition 22.3 The covering number of $\mathscr{F}$ at radius ε, with respect to d_n, denoted as $N(\mathscr{F}, \varepsilon, n)$ is the minimum size of function set that can cover $\mathscr{F}$ at radius ε.

According to the relationship between $\mathscr{F}$ and $\mathscr{G}$, it is not difficult to obtain that $N(\mathscr{F}, \varepsilon, n) = N(\mathscr{G}, \varepsilon, n)$.

When the covering number of a function class is finite, one can approximate the class by a finite set of functions that cover the class. In this way, we can make use of the finite union bound. The following results have been proven.

Theorem 22.4 $\forall t > 0$,

$$P\big(\exists g \in \mathscr{G} : R(g) > \hat{R}(g) + t\big) \leq 8E\big[N(\mathscr{G}, t, n)\big] e^{-nt^2/128}.$$

Actually, the covering number is not difficult to compute for many function classes. For example, if $\mathscr{G}$ is a compact set in a d-dimensional Euclidean space, we have $N(\mathscr{G}, \varepsilon, n) \approx \varepsilon^{-d}$. For a function class $\mathscr{G}$ whose VC dimension is h, we have $N(\mathscr{G}, \varepsilon, n) \leq Ch(4e)^h \varepsilon^{-h}$.

22.3.2.5 Uniform Bounds Based on Rademacher Average

To facilitate the concept of Rademacher average, we need to introduce the concept of Rademacher variables first. Rademacher variables are independent random variables $\sigma_1, \ldots, \sigma_n$ with $P(\sigma_i = 1) = P(\sigma_i = -1) = 1/2$. We denote E_σ as the expectation taken with respect to the Rademacher variables (i.e., conditionally on the data) while E as the expectation with respect to all the random variables. Then we have the following definitions.

Definition 22.4 For a function class $\mathscr{F}$, the Rademacher average is defined as

$$\mathscr{R}(\mathscr{F}) = E\left[\sup_{f \in \mathscr{F}} \frac{1}{n} \sum_{i=1}^{n} \sigma_i f(Z_i) \right]$$

and the conditional Rademacher average is defined as

$$\mathscr{R}_n(\mathscr{F}) = E_\sigma \left[\sup_{f \in \mathscr{F}} \frac{1}{n} \sum_{i=1}^{n} \sigma_i f(Z_i) \right].$$

The intuitive explanation of the Rademacher average is that it measures how much the function class can fit random noise. According to the relationship between $\mathscr{F}$ and $\mathscr{G}$, it is not difficult to obtain that $\mathscr{R}_n(\mathscr{F}) = \frac{1}{2} \mathscr{R}_n(\mathscr{G})$.

With the definition of the Rademacher average, one has the following theorem.

Theorem 22.5 $\forall 0 < \delta < 1$, *with probability at least* $1 - \delta$,

$$R(g) \leq \hat{R}(g) + 2\mathscr{R}(\mathscr{G}) + \sqrt{\frac{\log \frac{1}{\delta}}{2n}}$$

and also with probability at least $1 - \delta$,

$$R(g) \leq \hat{R}(g) + 2\mathscr{R}_n(\mathscr{G}) + \sqrt{\frac{2\log \frac{2}{\delta}}{n}}.$$

Similar to the covering number, it is not difficult to compute the upper bounds for the Rademacher average in many cases. For example, for a finite set with $|\mathscr{G}| = N$, we have $\mathscr{R}_n(\mathscr{G}) \leq 2\sqrt{\log N/n}$. And for a function class whose VC dimension is h, we have $\mathscr{R}_n(\mathscr{G}) \leq C\sqrt{h/n}$.

References

1. Bishop, C.M.: Pattern Recognition and Machine Learning. Springer, Berlin (2006)
2. Bousquet, O., Boucheron, S., Lugosi, G.: Introduction to statistical learning theory. In: Advanced Lectures on Machine Learning, pp. 169–207. Springer, Berlin (2004)
3. Mitchell, T.: Machine Learning. McGraw-Hill, New York (1997)
4. Ng, A.: Lecture notes for machine learning. Stanford University cs229 (2010). http://see.stanford.edu/see/courseinfo.aspx?coll=348ca38a-3a6d-4052-937d-cb017338d7b1s
5. Vapnik, V.N.: Statistical Learning Theory. Wiley-Interscience, New York (1998)

Index

T.-Y. Liu, *Learning to Rank for Information Retrieval*,
DOI 10.1007/978-3-642-14267-3, © Springer-Verlag Berlin Heidelberg 2011